Mathematics Education for Sustainable Economic Growth and Job Creation

Mathematics Education for Sustainable Economic Growth and Job Creation considers the need for young employees to be capable and confident with transferable knowledge and skills in mathematics and statistics in order to support economic growth in developing countries in an increasingly digital age.

This book draws on differing international perspectives in relation to mathematics education for sustainable economic growth and job creation. The contributors include education researchers and those involved in policymaking for both developing countries and beyond. Within each chapter, there is a reflection from the authors on their experiences in educational systems and policy development or research studies, which contribute to sustainable economic growth in different countries. As well as considerations of economies and job creation, the scholarship delves further into developing a critically aware citizenship through mathematics education.

Extending current thinking about the role of mathematics education and educating students for future needs, this book will be of great interest for academics, researchers and postgraduate students in the field of mathematics education, STEM education and sustainability education.

David Burghes is director of the Centre for Innovation in Mathematics Teaching (CIMT) in the Institute of Education, University of Plymouth, UK.

Jodie Hunter is an Associate Professor in the Institute of Education at Massey University, New Zealand.

Routledge Research in STEM Education

The *Routledge Research in STEM Education* series is home to cutting-edge, upper-level scholarly studies and edited collections covering STEM education.

Considering science, technology, engineering and mathematics, texts address a broad range of topics including pedagogy, curriculum, policy, teacher education and the promotion of diversity within STEM programmes.

Titles offer dynamic interventions into established subjects and innovative studies on emerging topics.

An Asset-Based Approach to Advancing Latina Students in STEM
Increasing Resilience, Participation, and Success
Edited by Elsa M. Gonzalez, Frank Fernandez and Miranda Wilson

Teacher Education to Enhance Diversity in STEM
Applying a Critical Postmodern Science Pedagogy
A. Anthony Ash II, Greg A. Wiggan and Marcia J. Watson-Vandiver

Teaching Early Algebra through Example-based Problem Solving
Insights from Chinese and U.S. Elementary Classrooms
Meixia Ding

Gender Equity in STEM in Higher Education
International Perspectives on Policy, Institutional Culture, and Individual Choice
Edited by Hyun Kyoung Ro, Frank Fernandez and Elizabeth Ramon

Mathematics Education for Sustainable Economic Growth and Job Creation
Edited by David Burghes and Jodie Hunter

For more information about this series, please visit: www.routledge.com/ Routledge-Research-in-STEM-Education/book-series/RRSTEM

Mathematics Education for Sustainable Economic Growth and Job Creation

Edited by David Burghes and Jodie Hunter

Routledge
Taylor & Francis Group

LONDON AND NEW YORK

First published 2022
by Routledge
2 Park Square, Milton Park, Abingdon, Oxon OX14 4RN

and by Routledge
605 Third Avenue, New York, NY 10158

Routledge is an imprint of the Taylor & Francis Group, an informa business

British Library Cataloguing-in-Publication Data
A catalogue record for this book is available from the British Library

Library of Congress Cataloging-in-Publication Data
A catalog record has been requested for this book

ISBN: 978-0-367-50028-3 (hbk)
ISBN: 978-0-367-50030-6 (pbk)
ISBN: 978-1-003-04855-8 (ebk)

Typeset in Bembo
by KnowledgeWorks Global Ltd.

Contents

Figures

Tables

Contributors

David Burghes is director of the Centre for Innovation in Mathematics Teaching (CIMT) in the Institute of Education, University of Plymouth, UK. He has interests in all aspects of teaching and learning of Mathematics in all sectors of education and has led three international projects comparing progress in the Primary and Secondary sectors and comparison of methods of teacher training. This research and development has led to curriculum projects for Primary and Secondary Mathematics and a new context-driven course for post-16 Mathematics in Colleges and Schools. CIMT has recently developed a part-time online Master's degree in Mathematics Education.

Jodie Hunter is an Associate Professor in Mathematics Education at the Institute of Education at Massey University, New Zealand. She began her career as a primary teacher in New Zealand before working as a Research Fellow at the University of Plymouth in the United Kingdom. She completed her doctorate focused on developing early algebraic reasoning shortly after returning to New Zealand to work at Massey University. Recently Jodie was a Fulbright scholar at the University of Arizona studying the use of funds of knowledge models to accelerate the achievement of diverse students. Within New Zealand, Jodie co-leads a large-scale professional development and learning project, which focuses on developing culturally responsive teaching to address under-achievement in mathematics for Pāsifika and Māori students at low socio-economic schools.

Joy Baker-Gibson is a passionate and committed Mathematics Educator with 21 years of teaching experience. She holds a Master's Degree in Mathematics Education with distinction from the University of the West Indies – UWI (2013). She was awarded the top graduate student in that year. In addition, she has a Bachelor's Degree in Mathematics Education (Honours, 2003) from UWI. She also received the award for the most outstanding Mathematics student. Mrs Baker-Gibson is currently serving as the Dean of the Faculty of Science and Technology at The Mico

University College. She has also served as Head of the Mathematics Department at The Mico. Additionally, she worked at Manchester High School where she taught Mathematics and served in the capacity as Head of Department. Mrs Baker-Gibson taught Mathematics and Chemistry at her alma mater for two years. She is also a devoted Christian and an active member in her church.

Albert Benjamin has been a lecturer at The Mico University (Jamaica) in Science Teaching Methodologies and STEM Education since 1999. He is a graduate of The University of the West Indies (Jamaica) and the University of Sheffield UK where he completed his doctoral studies in teacher education. He is currently Vice President of Academic Affairs at The Mico University College and a member of the National STEM Task Force (Jamaica). He has done pioneering work in developing STEM curricula for higher education and teacher professional development and is presentl a peer review appointee for the Journal of Education & Development in the Caribbean (JEDIC). Dr Benjamin is presently engaged as STEM consultant with the Organisation of American States ITEN and also with the Inter-American Development Bank. His research interests include teacher professional identity and effective pedagogies in science teacher education.

Shandelene Binns-Thompson is a PhD candidate at the University of Plymouth, UK and a Math Educator with Durham Public Schools, NC, USA. She was previously a lecturer and Associate Director of the Centre of Excellence in Mathematics Teaching at The Mico University College, Kingston, Jamaica. Under her leadership, she partnered with Professor Burghes and introduced the CIMT-based Subject Knowledge Enhancement (SKE) Programme and Mathematics Enhancement Programme (MEP) to the Caribbean for the first time. She has been working with a number of primary schools in Jamaica, helping them to adapt and implement the best practices of mathematically high-performing countries.

Neville Davies has a BSc, MSc and PhD from the University of Nottingham. From 1972 to 2009, he taught and conducted research in statistics at Nottingham Trent University, with visiting positions at universities in the United States, Australia and New Zealand. He became Professor of Applied Statistics (1994) and Director of the Royal Statistical Society Centre for Statistical Education (RSSCSE, 1999) and has served on several RSS committees, national and professional advisory bodies. He moved with the RSSCSE to the University of Plymouth, UK (2009) and was awarded an Emeritus position in statistical education (2014). He is a Trustee of the journal Teaching Statistics

Janet Georgeson is Associate Professor of Education at the Plymouth Institute of Education, UK, with a professional background in teaching children with special educational needs in a range of secondary, primary and pre-school settings. She has carried national and international research into professional development for practitioners to support their understanding of relational pedagogy, children's attitudes to mathematics and the development of computational thinking. Her research is strongly influenced by sociocultural and activity theory, in particular when applied to organisational structure, interactional style and approaches to pedagogy.

Julia Hill is a PhD student studying at the University of Melbourne's Graduate School of Education, Australia. Julia's background is in Educational Psychology and she undertook a research Master's in New Zealand exploring the values espoused by culturally diverse mathematics learners. She continues to collaborate on research projects in New Zealand focusing on culturally responsive and equitable mathematics education. Julia's current research interests centre on understanding student well-being specific to mathematics education, and how students' values might relate to and impact on their well-being in mathematics education.

Roberta Hunter MNZM is a New Zealand education academic of Cook Islands Māori descent and as of 2019 is a full professor at the Massey University, New Zealand. She specialises in mathematics education. In the 2020 Queen's Birthday Honours, Roberta was appointed a Member of the New Zealand Order of Merit, for services to mathematics education.

Margaret L Kern is an associate professor at the Centre for Positive Psychology at the University of Melbourne's Graduate School of Education, Australia. Her research focuses on understanding, measuring, and supporting well-being across the lifespan. She works with schools and workplaces to examine strategies for supporting well-being, and bridging gaps between research and practice. She has published 3 books and over 100 peer-reviewed articles and chapters.

Masataka Koyama is a professor of mathematics education, Graduate School of Humanities and Social Sciences at Hiroshima University in Japan. He had been the dean of Graduate School of Education for four years from 2016 to 2020 at Hiroshima University. His major scholarly interests are mathematical understanding, international comparative study on students' mathematical attainments, mathematics teachers' professional development and school mathematics curricula and textbooks. He has been actively involved in both the international activities such as ICME, EARCOME and PME and the national activities for society of mathematics education and teachers' professional development through lesson study of mathematics in primary and secondary schools.

Leo Pahkin (MSc) is Counsellor of Education at the Finnish National Board of Education. His main area is the development of mathematics and science education in pre-school, basic and upper secondary general education. He has been a project manager for gifted and talent children in a national program. He has worked as a subject teacher (mathematics, physics, chemistry and ICT), teacher trainer and researcher for several years before this current post. He has written several articles in journals and he is also deputy member of the Finnish matriculation examination board and a member of the mathematics committee of European Schools.

Asburn Pinnock is President of The Mico University College, Jamaica. He was formerly Principal of the Sam Sharpe Teachers' College, Jamaica. He has a Doctorate of Education in Educational Leadership, having previously gained a Diploma in Teaching, specialising in History and Social Studies, and a Master of Science in International Area Studies.

Rachel Restani was recently a Postdoctoral Research Fellow for the Developing Mathematical Inquiry Communities (DMIC) project at Massey University, New Zealand, which aims to elevate Maori and Pāsifika mathematical success in Year 0–Year 10 classrooms through in-class support and PLD. Rachel has experience educating high school mathematics learners from ethnically and socioeconomically diverse backgrounds in California. Rachel worked as a researcher on two PLD projects aimed at supporting teachers to respond to students' ideas. She also supervised pre-service teachers and was a teaching assistant for postgraduate and undergraduate courses before working as a Research Fellow for Massey.

Derek Robinson is an inspirational teacher, leader and speaker in the United Kingdom. He has consistently researched 'good practice' from other countries, particularly from Hungary and Japan, which he has then trialled, developed and shared with many teachers from across England. He is one of the country's leading experts on Japanese Lesson Study and the Japanese Approach to teaching mathematics through problem solving and has been a consultant to CIMT (Centre for Innovation in Mathematics Teaching) at the University of Plymouth, UK. His leadership experience includes training and developing teaching personnel, curriculum writing and development, initiating and implementing alternative forms of assessment and evaluation.

Naomi Sani has worked in the education sector for 30 years, and has been involved with teaching, training and research in the primary, secondary and tertiary sectors. Naomi is currently a Lecturer in Education with University of Plymouth, UK and completed her PhD in Mathematics Education in 2019. Concurrently, Naomi was the 'mathematics expert' with the Education Development Trust (EDT) for the Teaching and Learning International Study (TALIS) Video Study project with the OECD (May 2016–November 2020). This successful and innovative project strived to push the frontiers of international education research, with many valuable methodological lessons learnt, and an array of interesting findings and resources generated. Naomi is the author of various papers and articles, and of the popular book 'How To Do Maths So Your Children Can Too'.

Padmanabhan Seshaiyer is a Professor of Mathematical Sciences at George Mason University, USA. He also serves as the director of the STEM Accelerator Program and the Center for Outreach in Mathematics Professional Learning and Educational Technology. His research interests are in the broad areas of mathematical modelling, computational thinking, data science, design thinking, entrepreneurship, UN Sustainable Development Goals and STEM education. During the last decade, he has initiated and directed a variety of educational programs to foster the interest of students and teachers in STEM at all levels. He is also actively involved in multiple STEM collaborative projects globally.

Wee Tiong Seah is Professor of Mathematics Education in the Melbourne Graduate School of Education at The University of Melbourne, Australia. He is passionate about harnessing conative variables to foster cognitive and affective development in learners. His research interests include values for – and competences/proficiencies through – mathematics pedagogy, as well as mathematical well-being. He embraces cross-cultural comparative research as a means of promoting insightful education reform and coordinates a 20-nation research consortium.

Jan van Driel is a Professor of Science Education and leader of the Mathematics, Science & Technology Education Group in the Melbourne Graduate School of Education at the University of Melbourne, Australia. His research interests include science teacher knowledge, teacher education and professional learning, science and gender and interdisciplinary science and STEM education. Currently, he is co-editor-in-Chief of the *International Journal of Science Education*.

Neville Ying is Professor and Pro-Chancellor of The Mico University College, Kingston, Jamaica. He has a Mathematics and Physics background and his PhD was in Measurement, Statistics and Research Design. He is a trained teacher and was instrumental in establishing the first Mathematics Summit in 2019. He is also Executive Director of the Jamaica Diaspora Institute and has won many awards including the Jamaica National Honours of Commander of Distinction.

Foreword

The Mico University College is proud to have partnered with the Ministry of Education, Youth and Information (MOYE) in hosting the Mathematics Summit under the theme 'Mathematics for Sustainable Economic Growth and Job Creation'. It is my pleasure and honour to write the foreword to this research monograph, based on papers presented at the summit.

Within the context of higher education and the Fourth Industrial Revolution, the process of teaching and learning continues to impact economic growth globally. It has been well documented that consistent and sustained economic growth has been eluding developing countries, including Jamaica, for many decades. Research has shown that there is a close relationship between education attainment (especially in Mathematics and Science) and economic growth. Educational institutions, especially those in higher education, are mandated to prepare a workforce of problem-solvers, creative and innovative thinkers and communicators.

Against this background, The Mico University College is pleased to have pioneered this initiative. Our focus on Mathematics for Sustainable Economic Growth and Job Creation for this summit puts into perspective the skills and competencies we believe the subject area fosters as foundation for other academic subjects. It is prudent to equip our student teachers, in-service teachers and Mathematics tutors with the best practices from top performing countries around the world.

The summit was a game changer for Jamaica and the Caribbean. The knowledge-sharing forum provided opportunities to explore, analyse and consolidate international best practices in the teaching of Mathematics to:

- unleash creativity;
- address misconceptions in the subject of mathematics;
- increase proficiency in problem solving and
- influence policies and systems in the education sector.

The three days of the summit allowed for stimulating and informative discussion between delegates from the participating countries and gave opportunities for an exchange of knowledge and experiences in education that will be of benefit to all the participants.

Chapters included in this monograph build on the discussions at the summit, helping developing countries to harness the potential of mathematics and data science to improve and sustain economic growth and job creation.

Dr Asburn Pinnock,
JP, President,
The Mico University College, Jamaica

Acknowledgements

With sincere thanks to our colleague, Liz Holland, whose work has helped to ensure that these chapters are coherent, consistent and clear.

Chapter 1

Mathematics education for sustainable economic growth and job creation: Setting the scene

David Burghes, Jodie Hunter and Neville Ying

Background

This edited book gathers a collection of work from around the world to consider a range of perspectives on how mathematics education can be used as a vehicle to promote sustainable economic growth and job creation. The papers are based on talks and discussion at the first Mathematics Summit held in Kingston, Jamaica in March 2019.

This opening chapter considers the rapid changes in technology across the world that have led to questions related to the type of mathematics education that will prepare students to be productive citizens in societies of the future. The chapter describes the context and background related to the focus of the book and argues for the need to consider pedagogical practices, curriculum content, professional learning and development when reforming mathematics education to align with 21st-century skill requirements and technological advances. Finally, it introduces the content of the chapters and the contributions that each of these makes.

Context

For countries all over the world, a key question is how to shape education systems that are future-ready and adequately prepare individuals to contribute in a variety of ways. When we consider mathematics education that promotes sustainable economic growth and job creation, we need to be aware of a range of factors. As previously asked in a journal article by Gravemeijer et al. (2017, p. 105) for those working in mathematics education, this includes consideration of what 'may prepare students for the society of the future'. While there is a need to think about the role of technology, we must also consider the skills that students need to develop, along with key competencies that will equip them for future work and to be responsible, socially aware citizens. To achieve these aims, we need to consider both the content of the curriculum used in our mathematic classrooms and the ways in which we teach. Importantly, how can we reform mathematics education to support the development of 21st-century skills and what does this mean for developing countries?

A changing world: Technological advances, creating socially aware citizens and reforming mathematics education

Mathematics has become increasingly important across society and in the workforce; there are also significant changes in the way that mathematics is used. In many fields or work, there has been increasing automation which means that mathematics is completed by machines and hidden from view (Gravemeijer et al., 2017). In addition to automation, we are seeing growing digitalization and new and emerging technology such as artificial intelligence, big data, block chain, splinternets and nanotechnology. These rapid changes and developments raise important questions for every country in relation to the forms of education that will be needed to support their society to prosper and grow (Beswick & Fraser, 2019; Fullan, 2001). This is particularly relevant for developing countries such as Jamaica but these changes, while challenging, can also be viewed as an opportunity for development and reform of educational systems.

Teaching mathematics for sustainable economic growth and job creation requires more than just preparing students for the workforce; it also encompasses education systems that prepare individuals for active socially responsible citizenship (Gravemeijer et al., 2017; Maass, Doorman, Jonker, & Wijers, 2019). This includes developing quantitative literacy defined by Simic-Muller (2019) as 'the ability to see the world through a mathematical lens'. Encompassed within this is being able to use mathematics to recognize systemic injustice related to areas such as wealth or ethnicity and focus on developing social-justice-oriented solutions. Reforming mathematics education to grow socially responsible citizens and to align with 21st-century skills and proficiencies useful for developing sustainable economic growth requires reflection on both the content of the mathematics curriculum and the types of pedagogy which are used.

Arguably, in many countries there is a contrast and a gap between the mathematics curricula taught within schools and what is useful within the workforce. One way to bridge this gap is to further develop mathematics curricula and tasks to include messy real-world situations which require the use of mathematical modelling to develop possible solutions (Maass et al., 2019; Simic-Muller, 2019). This in itself presents challenges: for example, Simic-Muller (2019) identifies the time-consuming nature of developing such curricula, difficulties in connecting specific topics to real-world issues, and maintaining authenticity of the context while also engaging with mathematics. However, as shown in research studies (e.g., Maass et al., 2019; Simic-Muller, 2019) the use of such tasks has the potential to enhance students' mathematics reasoning skills while also transforming their perceptions in relation to the usefulness of mathematics and how it is relevant to their communities.

When introducing new types of curricula and enabling students to understand mathematics differently and to develop a range of competencies that complement the role of technology, we need to reflect on the type of pedagogical practices that will support this. A number of researchers (e.g., Beswick & Fraser, 2019; Gravemeijer et al., 2017; Maass et al., 2019; Teo, 2019) have identified the key competencies that are required for the future: these include expert communication skills, critical thinking, creativity and collaboration. Research studies (Maass et al., 2019; Simic-Muller, 2019) drawing on modelling tasks or inquiry-based learning, illustrate how students can develop their communication skills and critical thinking while engaging in mathematical tasks. Key aspects of instruction that appear to support the development of key competencies are: focusing on the strengths that students bring to the task; providing opportunities for discussion and communication; allowing students to engage in critical thinking through considering opinions as well as evidence and making choices.

In this edited book, we draw together a range of elements that contribute to the development of mathematics education systems that support countries in relation to sustainable economic growth and job creation. The chapters span mathematics education in relation to one or more of the following inter-related aspects:

1 Pedagogical practices: The type of pedagogical practices that can be used by educators to support students to develop 21st-century skills and key competencies including critical thinking, communication and problem solving;
2 Curriculum content: The content of the mathematics curriculum that will facilitate students to develop deep understanding of mathematics and recognize mathematics in context. This includes supporting students to recognize mathematical problems in the real world;
3 Professional learning and development (PLD): Designing PLD that will have a transformational impact on teaching and learning in classrooms.

Overview of the book

This book draws on differing international perspectives in relation to mathematics education for sustainable economic growth and job creation. It includes contributions from education researchers and those involved in policy making for both developing countries and beyond. Within each chapter is a reflection from the authors on their experiences in educational systems and policy development or research studies which contribute to sustainable economic growth in different countries. This extends beyond simply economics and job creation into mathematics education which supports the development of contributing and critically aware citizenship. It is intended that this book will extend current thinking about the role of mathematics education in relation to growing the economy of a country and educating students for societies of the future.

The book begins with two chapters to provide a background to the overall focus on mathematics education for sustainable economic growth and job creation. This includes considering why this has become an area of growing importance due to changes in technology and the workplace. An overview of the benchmarks in international studies related to mathematics education is presented (Chapter 2). The subsequent two chapters illustrate two case studies of countries identified as high performing in international studies related to mathematics. The first of these chapters reflects on the policies and programmes for innovation in Finland (Chapter 3) while the second considers the common teaching approaches used in Japan and how the professional learning and development practice of lesson study has contributed to improving performance with regard to both teaching and learning (Chapter 4).

Chapters 5–7 draw on a range of international research to consider innovations in mathematics teaching and learning for developing countries. As a collective, these chapters highlight a range of factors to be considered when reflecting on mathematics education for sustainable economic growth and job creation. At the core of the first two chapters in this section is the need for equity for diverse learners. Chapter 5 highlights the rich mathematical contexts of students and their families in a small island situation. This links to supporting students to recognize mathematics in their own contexts. Chapter 6 uses an example of Pāsifika students in New Zealand to demonstrate how best practice and acceleration of academic achievement can be accomplished by reforming pedagogical practices to draw upon these students' cultural beliefs and values. The following chapter looks beyond a focus on cognitive aspects and academic achievement with regard to developing citizenship to contribute to the society of the future. This broadened perspective of important elements within the mathematics classroom includes an explicit focus on well-being for diverse learners (Chapter 7).

The next two chapters focus on aspects of professional learning and development which support mathematics educators in reforming their pedagogical practices to align with the development of rich mathematical understanding and 21st-century skills and proficiencies. Chapter 8 outlines novel education frameworks that educators can draw upon to prepare students to become life-long learners who can then go on to pursue state-of-the-art employment. Chapter 9 reports on the impact of providing a subject knowledge enhancement platform for teachers of mathematics. The final chapter in this section considers science, technology, engineering, and mathematics (STEM) as a key aspect of mathematics education to support 21st-century economies. Of particular interest in this chapter are the implications of the STEM agenda for developing countries (Chapter 10).

The final section of the book advocates the priorities for future action to reform education systems. This includes developing 'applied' courses which focus on problem solving as a key competency, and introducing a new curriculum based on Data Science. In Chapter 11, more details about teaching through problem solving and the use of lesson study are considered and how this can enhance professional development in teaching. The following chapter (Chapter 12) describes a case study from England of 'Core Maths' courses to support learners to use mathematical and statistical skills in new areas of application. Similarly, Chapter 13 makes a case for the introduction of Data Science in schooling and the development of teaching and learning resources to align with a STEM agenda and upskill local populations.

Finally, Chapter 14 provides a summary of the evidence presented, with suggestions and recommendations put forward in the previous chapters.

References

Beswick, K. & Fraser, S. (2019). Developing mathematics teachers' 21st century competence for teaching in STEM contexts. *ZDM, 51*, 955–965. https://doi.org/10.1007/s11858-019-01084-2

Fullan, M. (2001). *The new meaning of education change*. New York, Teachers College Press.

Gravemeijer, K., Stephan, M., Julie, C., Lin, F., & Ohtani, M. (2017). What mathematics education may prepare students for the society of the future? *International Journal of Science and Mathematics Education, 15*, 105–123.

Maass, K., Doorman, M., Jonker, V., & Wijers, M. (2019). Promoting active citizenship in mathematics teaching. *ZDM, 51*, 991–1003. https://doi.org/10.1007/s11858-019-01048-6

Simic-Muller, K. (2019). "There are different ways you can be good at math": Quantitative literacy, mathematical modeling and reading the world. *PRIMUS, 29*(3-4), 259–280.

Teo, P. (2019). Teaching for the 21st century: A case for dialogic pedagogy. *Learning, Culture and Social Interaction, 21*, 170–178.

Chapter 2

Incentives for enhancing mathematical literacy in developing countries

Neville Davies and Janet Georgeson

Background

Dr Alfonso Echazarra, Director for Education and Skills, Organisation for Economic Co-operation and Development (OECD) presented some of the material in this chapter, in two presentations at the first International Mathematics Teaching Summit, held in Kingston, Jamaica in March 2019. The presentations were entitled:

- *The Role of PISA: Aligning Jamaica and the Caribbean with International Benchmarks for Mathematics;*
- *Jamaica in PISA.*

Dr Echazarra presented the justification for using the Programme for International Student Assessment (PISA) tests for international benchmarking, in all countries, including those characterised as 'developing'. He outlined by data summaries and graphs (reproduced here), mathematical competence but also information from questionnaires on teaching and learning strategies and their relationship with scores in PISA tests. We are grateful to Dr Echazarra for permission to use the content of his PowerPoint presentations.

We add to Dr Echazarra's material by considering, amongst other things, the implications that result from the criticisms statisticians have made concerning the flaws in the modelling process followed by the OECD using the PISA data. The statistical issues imply that the rankings that result from ordering the PISA mathematics scores have more variability and uncertainty than is apparent by just looking at the raw data, or even the confidence intervals the OECD gives.

In spite of statistical problems, we suggest there may be merits for countries like Jamaica in considering what other countries have done, because of PISA, in order to enhance their teaching and learning strategies in mathematics.

Setting the scene

In this chapter, we first consider the role of PISA, run by OECD, in providing incentives for teachers and policy makers in mathematics in developing countries such as Jamaica to help improve mathematical literacy. Started in 2000 and run every three years, PISA creates league tables of countries' performance in mathematics and other subjects, as measured by a test taken by young students in schools. For mathematics, the OECD defines mathematics literacy as follows:

> An individuals' capacity to formulate, employ, and interpret mathematics in a variety of contexts. It includes reasoning mathematically and using mathematical concepts, procedures, facts and tools to describe, explain and predict phenomena. It assists individuals in recognising the role that mathematics plays in the world and to make the well-founded judgements and decisions needed by constructive, engaged and reflective citizens.

This statement forms the background to the mathematics tests devised by the OECD.

This view of mathematical literacy refers to the role of the subject in the real world and the ways in which it can help in decision-making. Unfortunately, it does not recognise the role mathematics plays in statistics and Data Science (DS), the science of learning from data. DS, which many see as a development of the subject *statistics*, plays a key role in understanding and managing the world around us. This is discussed in more detail in a later chapter where we also highlight what it means to be *statistically* literate. We argue that it would be useful if the OECD recognised the emergence of DS, the ways in which it can be taught, and its potential in school-level education, either within the remit of mathematics or as a discipline in its own right.

Historically, many schoolteachers, at least, regard statistics as a branch of mathematics. In any case, many mathematics curricula contain aspects of statistics and its applications. The more progressive ones teach *statistical problem solving* as a way for school students to explore uncertainty and decision making in the real world. In the later chapter, we consider DS, statistical problem solving and learning from data and argue that teachers of mathematics in countries characterised as 'developing' should give their students experience of solving real-world problems through embedding these topics in their own curricular.

At the time of the conference (March 2019), the 2015 PISA results (published in 2016) were the most recent ones available. The results for 2018 were published in December 2019 and we will refer to the newer results when appropriate. The 2018 results included analysis of data for around 600,000 participating students in 79 countries and economies. China's economic areas of Beijing, Shanghai, Jiangsu and Zhejiang emerged as the top performers in all categories: mathematics; reading and science. For a full listing of each country's mathematics rankings from 2000 to 2019 see: https://en.wikipedia. org/wiki/Programme_for_International_Student_Assessment

Brief history of PISA, benchmarking and improving quality of teaching

PISA is a test in mathematics, reading and science taken by 15-year-old students. When it started in 2000 there were 40 participating countries but has gradually gained in popularity increasing steadily to 79 countries in 2018. The examination comprises a computer-based test lasting a total of 2 hours. Table 2.1 shows year of implementation of PISA and the domains tested.

It usually takes a year before OECD official reports are published and so, for example, the 2018 tests were published in 2019. There are certain options of financial literacy (started in 2012) and foreign language (starts in 2024). Other domains include: problem solving; collaborative problem solving; creativity and digital reading.

The marks attained by the students are used as a measurement of the success of that country in providing, for example, mathematics education at that level. In 2015, more than half a million students represented more than 30 million 15-year-olds in the schools of the participating countries and economies. In practice, not all the students answer all the questions posed and this results in statistical issues related to what to do about the missing vales in the responses (see Goldstein, Carpenter, & Browne, 2014). The estimated overall scores in the PISA exercise uses Item Response Theory (IRT). This is a paradigm for the design, analysis, and scoring of tests, questionnaires and similar instruments measuring abilities, attitudes or other variables.

Recent PISA tests not only evaluate if students can reproduce what they have learned at school but also assess their capacity to apply creatively their knowledge and skills in a variety of situations related to science, reading, mathematics, financial literacy, problem solving and other domains.

They answer questions about their schools, personal context and attitudes towards learning. Parents, principals, teachers and policy-makers provide information about school policies, practices, resources and institutional factors that can explain the differences in performance.

In terms of raw score rankings in PISA 2015, the top seven countries (Singapore, Hong Kong (China), Macao (China), Chinese Taipei, Japan, B-S-J-G (China) and Korea) are in the Far East, with New Zealand, Australia and the United Kingdom in 21st, 25th and 27th positions, respectively. The PISA reports break down the rankings into three broad bands: average, above and below, and some academics think that this is as detailed

Table 2.1 PISA cycles and cognitive domains

Year	2000	2003	2006	2009	2012	2015	2018	2021
Domains	Reading	Reading	Reading	Reading	Reading	Reading	Reading	Reading
	Maths	Maths	Maths	Maths	Maths	Maths	Maths	Maths
	Science	Science	Science	Science	Science	Science	Science	Science
		Problem solving			Problem solving	Collaborative problem solving	Global competency	Creativity
					Financial literacy	Financial literacy	Student well-being	Teachers' well-being

as any comparisons between countries should get. Unfortunately, the raw rankings can lead some commentators in the media to classify countries as being 'failures' if they are below the Far East countries or indeed the OECD average score of 487.

The first results published in 2001 sparked heated debate: the education landscape revealed by the assessment results was very different from what many had thought they knew. The data and the modelling the OECD also gave rise to several statistical criticisms. We shall describe some of the issues later in this chapter.

Enthusiasts promote the results of PISA as a tool to help countries to learn and improve, because it can engender collaboration between countries, experts and social agents by sharing experiences, policies and best practices. Countries can monitor progress and can help establish national objectives for education policy (e.g., mean performance, coverage, equity, low achievement, top performers, grade repetition, gender gap, well-being). Some countries have shown that success can become a consistent and predictable education outcome, and results have also shown that there is no automatic link between social disadvantage and poor performance in school.

One of the most important suggestions, although sometimes clouded in controversy, is that insights from several cycles of PISA can change and improve education systems. One reason why this causes controversy is because the data are not longitudinal and analysis requires this.

There were six levels of proficiency in mathematics in PISA 2015. Table 2.2 gives details of what students at each of the levels of proficiency should have been able to demonstrate.

Students who achieved levels 5 and 6 are classified as *top performers*, while levels 2 and 1 correspond to *low achievers*. The following two charts show the percentage of the top performers and low achievers in countries from PISA 2015. Figure 2.1 shows the ranking of 69 countries according to the percentage of top performers.

From this chart, the OECD average percentage of students who achieve a top performance proficiency of 5 and above in mathematics is just over 10%. This can be used as a benchmark for countries to achieve, if they are below the score. There were 23 countries where students achieved above this average and 46 countries below.

Figure 2.2 shows the ranking of 69 countries according to the percentage of low achievers. Their scores range from 6% (Macao) to 90% (Dominican Republic) with the OECD average being 23%.

Figures 2.1 and 2.2 show that many countries that are in the lowest 20 of high achievers are also in the top 20 of low achievers. The reasons for this will depend upon many factors, including methods of teaching, time spent on PISA-related topics, the relative importance ascribed to those topics and so forth.

Table 2.2 Six levels of mathematics proficiency in PISA 2015

Levels	What students can typically do
6	At Level 6, students can conceptualise, generalise and utilise information based on their investigations and modelling of complex problem situations, and can use their knowledge in relatively non-standard contexts. They can link different information sources and representations and flexibly translate among them. Students at this level are capable of advanced mathematical thinking and reasoning. These students can apply this insight and understanding, along with a mastery of symbolic and formal mathematical operations and relationships, to develop new approaches and strategies for attacking novel situations. Students at this level can reflect on their actions, and can formulate and precisely communicate their actions and reflections regarding their findings, interpretations, arguments and the appropriateness of these to the original situation.
5	At Level 5, students can develop and work with models for complex situations, identifying constraints and specifying assumptions. They can select, compare and evaluate appropriate problem-solving strategies for dealing with complex problems related to these models. Students at this level can work strategically using broad, well-developed thinking and reasoning skills, appropriate linked representations, symbolic and formal characterisations and insight pertaining to these situations. They begin to reflect on their work and can formulate and communicate their interpretations and reasoning.
4	At Level 4, students can work effectively with explicit models for complex concrete situations that may involve constraints or call for making assumptions. They can select and integrate different representations, including symbolic, linking them directly to aspects of real-world situations. Students at this level can utilise their limited range of skills and can reason with some insight, in straightforward contexts. They can construct and communicate explanations and arguments based on their interpretations, arguments and actions.
3	At Level 3, students can execute clearly described procedures, including those that require sequential decisions. Their interpretations are sufficiently sound to be a base for building a simple model or for selecting and applying simple problem-solving strategies. Students at this level can interpret and use representations based on different information sources and reason directly from them. They typically show some ability to handle percentages, fractions and decimal numbers, and to work with proportional relationships. Their solutions reflect that they have engaged in basic interpretation and reasoning.
2	At Level 2, students can interpret and recognise situations in contexts that require no more than direct inference. They can extract relevant information from a single source and make use of a single representational mode. Students at this level can employ basic algorithms, formulae, procedures or conventions to solve problems involving whole numbers. They are capable of making literal interpretations of the results.
1	At Level 1, students can answer questions involving familiar contexts where all relevant information is present and the questions are clearly defined. They are able to identify information and to carry out routine procedures according to direct instructions in explicit situations. They can perform actions that are almost always obvious and follow immediately from the given stimuli.

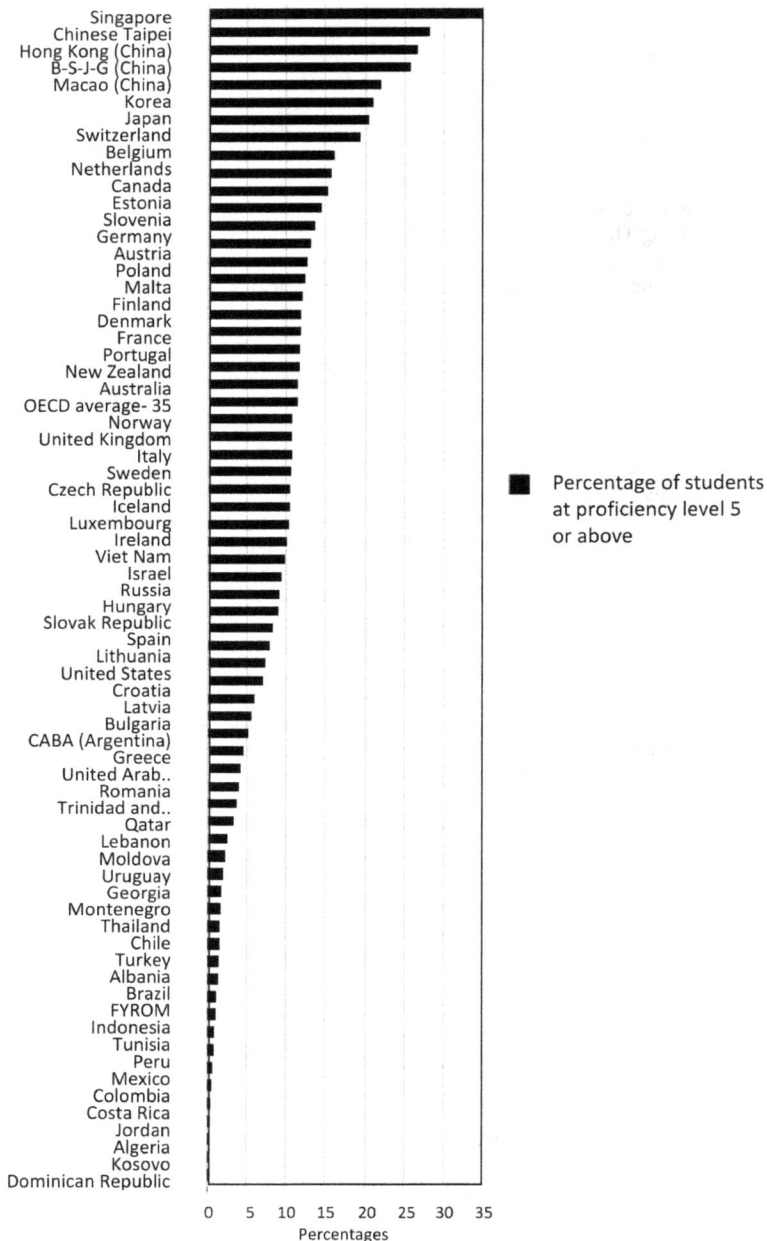

Figure 2.1 Percentage of top achievers in mathematics.

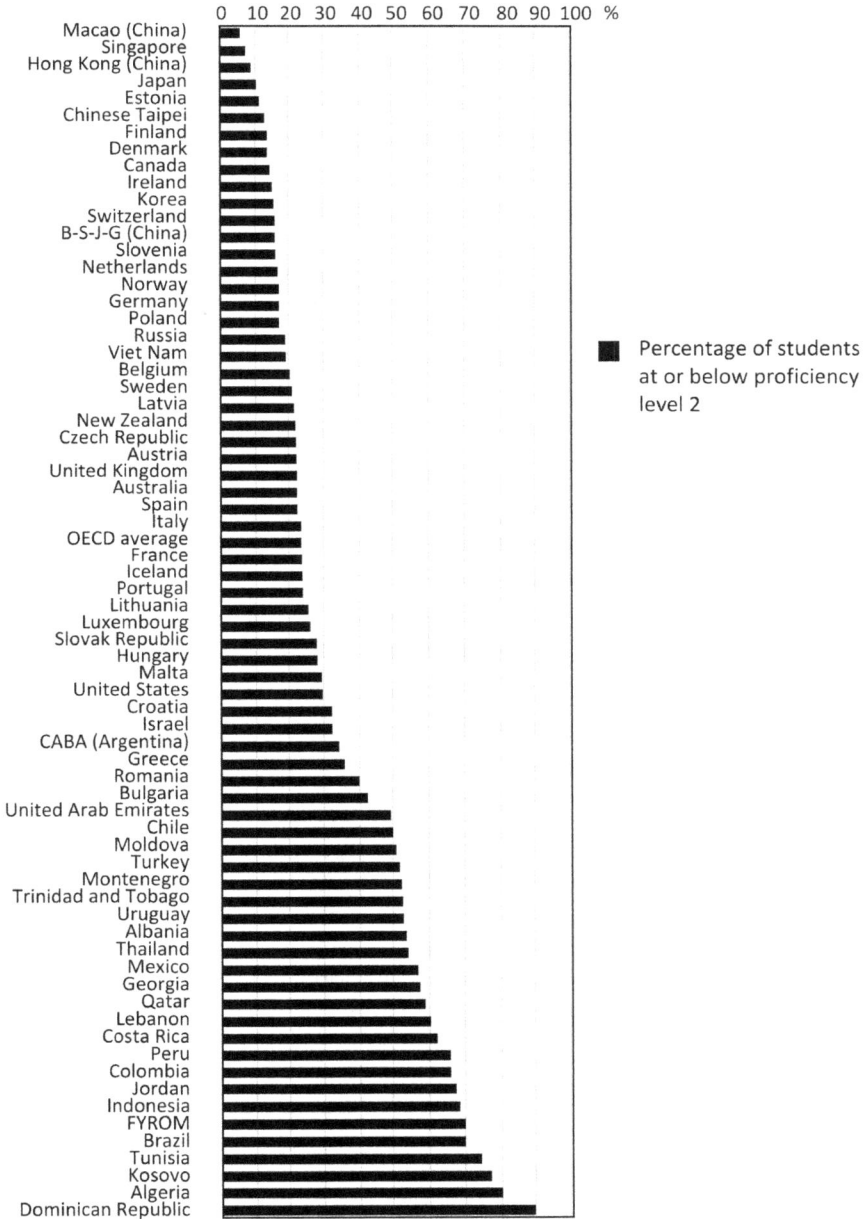

Figure 2.2 Percentage of low achievers in mathematics.

Other charts compare boys' and girls' scores and show that boys perform better in the PISA mathematics test when compared with girls. The top ten countries for boys' performance are:

> Austria, Lebanon, Argentina, Italy, Chile, Germany, Costa Rica, Ireland, Spain and Brazil.

Countries where girls perform better are:

> Trinidad and Tobago, Jordan, Georgia, Qatar, Albania, Macao (China), Finland, Korea, Algeria and United Arab Emirates.

There is also evidence that, on average, girls are more anxious about mathematics than boys across all countries. The reasons for this need to be investigated, but it is worth noting that PISA reports a gap of (15–23)% between the anxiety expressed by girls and boys for the countries: Liechtenstein (23%), Finland, Denmark, Shanghai, France, the United Kingdom, Switzerland, Macao, Germany, Canada, Luxembourg, Hong Kong and New Zealand (15%). It would be useful to find out whether differences in levels of anxiety exist between boys and girls in countries classified as 'developing'.

The experience of middle- and low-income countries in PISA

PISA results highlight differences in education quality between high- and middle-income countries. Students in middle-income countries generally perform at the lower levels of the PISA proficiency scales. If we separate middle- and low-income countries, they tend to perform at or below level 2.

However, some of the contextual factors measured by PISA are different in middle- and high-income countries. Because out-of-school rates are high in middle-income countries, indices of coverage can be as low as 50%.

Some countries report successful policy changes because of the results of PISA. For example, Brazil has pegged education expenditures to resource revenues, and reduced school-entry age; in the early 2000s, Turkey extended compulsory schooling and improved resource allocation; Portugal reduced grade repetition, consolidated schools and invested in teacher professionalism; finally France also reduced grade repetition.

Teaching and learning strategies

Teaching and learning strategies in mathematics vary greatly around the world. For example, according to the OECD, teaching in the United Kingdom, Australia, New Zealand, Ireland, Uruguay and Israel encourages much more memorisation than elaboration. In France, Hungary, Shanghai and Ireland teacher-directed instruction is employed much more than student-oriented instruction. On the PISA scale, when the difficulty of mathematics items is harder, it appears that memorisation is associated with a lower chance of success.

Furthermore, elaboration strategies can be associated with a greater chance of success as problems become more difficult. The students who are most successful at solving harder problems benefit from a combination of elaboration strategies and strategies that support them to take control of their own learning.

It has long been known that a range of cognitive skills contribute to success in learning many subjects (see, e.g., Anderson & Krathwohl, 2001; Bloom et al., 1956), and when at least some of those skills are activated in a mathematics teaching context, students seem to perform better in the PISA tests. For example, Marriott, Davies, and Gibson (2009) showed that a surprisingly wide range of cognitive skills, as defined by the modification of Bloom's taxonomy of learning by Anderson and Krathwohl (2001), are needed to successfully complete the statistical problem solving cycle.

Since problem solving in mathematics has become a key feature of the PISA tests, it would be useful to devise a research project to identify those cognitive domains that are needed in various stages of solving *mathematical* problems.

Posing a range of questions that focus on different cognitive domains, might improve cognitive skills that, in turn, provide answers to help solve problems.

Issues with PISA mathematics performance rankings

Since the first implementation of PISA the statistical methodology implicit in the implementation and analysis of the mathematics tests has come under scrutiny. Goldstein (2004a, 2004b, 2019) comments on the restricted nature of the data modelling and analysis, and the resulting interpretations. He points to certain features of the results that raise questions about the adequacy of the data and he stresses the failure to introduce a longitudinal component (see Setting the scene). Goldstein's papers make suggestions for ways in which such studies could be improved. Indeed, there are key methodological health warnings that other eminent statisticians have highlighted about the conduct of PISA tests, their analysis and the resulting published scores. Hopfenbeck et al. (2018) reviewed PISA-related English-language peer-reviewed articles from the programme's first cycle in 2000 to 2015 inclusive. Although, their findings indicated that studies based on the PISA datasets have led to progress in educational research, at the same time they urge caution when using this research to inform educational policy.

Policy makers can get 'hung up' on their country's league position from one PISA exercise to the next. For example, in 2009 New Zealand politicians seriously overreacted to a drop by a few places in their mathematics position and the words 'nosedive', 'plummet' and 'slump' were splashed across newspaper headlines. In response, the New Zealand government spent a lot of money sending people to Asia to find out what New Zealand mathematics teachers were doing wrong.

Of course, statistically, New Zealand and the surrounding countries in the rankings were not any different when you take uncertainty in estimates of league positions into account. Even when countries rank highly in PISA, as New Zealand has every time (until, arguably, in 2009) politicians never stop coveting the next rung on the ranking ladder.

Goldstein and Spiegelhalter (1996) and Goldstein (1997) first investigated issues with using ranking league tables' indicators of educational performance. PISA provides 95% confidence intervals for each country's score in mathematics. However, because of the inadequacies in the modelling carried out to produce the intervals, they are gross underestimates of what they should be. For example, Spiegelhalter (2013) provided a statistical argument that called into doubt the trustworthiness of the entire modelling procedure used by the OECD. He lists five related issues:

1 Individual students only answer a minority of questions, creating missing values in responses.
2 Multiple imputations (called 'plausible values' by OECD) are generated for all students with omitted answers.
3 'Plausible values' are then treated as if the students provided the answers. They then form the basis of national scores and hence rankings in league tables.
4 The statistical model used to generate the 'plausible scores' is inadequate.
5 This means the uncertainty in the national scores are underestimated and so the rankings are unreliable

Spiegelhalter comments:

PISA does have a lot to offer, but as a statistician I'm left with serious concerns about the reliability of the league tables and the lack of evidence to support PISA's methodology in compiling them. Governance and external review seem extremely limited. I will treat the scores and ranks with considerable skepticism.

For further details, see https://www.statslife.org.uk/features/1074-the-problems-with-pisa-statistical-methods

Notwithstanding these comments, there are some useful results and experiences that countries such as Jamaica may wish to consider. In the next section, we provide some details that teachers and policy makers might like to examine.

PISA in Jamaica and developing countries

The previous section included some evidence that the rankings implied by the mathematics scores achieved by students could be misleading. On the other hand, taking part in PISA exercises offers the potential to provide incentives for teacher improvement through examining what other countries do. Whether or not developing countries should enter PISA needs therefore to be given careful thought.

Two freely available publications from the Organisation for Economic Co-operation and Development, OECD (2016a,b) present some incentives and guidelines for mathematics teachers to follow, which could be useful to countries seeking to develop mathematical literacy. OECD (2016a) is about 'How Teachers Teach and Students Learn: Successful strategies for School', a PISA an educational working paper that:

> ... examines how particular teaching and learning strategies are related to student performance on specific PISA test questions, particularly mathematics questions. The report compares teacher-directed instruction and memorisation learning strategies, at the traditional ends of the teaching and learning spectrums, and student-oriented instruction and elaboration learning strategies, at the opposite ends.

See https://www.oecd-ilibrary.org/education/how-teachers-teach-and-students-learn_5jm29kpt0xxx-en

OECD (2016b) analyses the PISA student background questionnaire about students' experiences in their mathematics classes, including their learning strategies and the teaching practices they said their teachers used. The report, using data from the 2012 PISA exercise, and The Teaching and Learning International Survey (TALIS) of 2013,

> ... takes the findings from these analyses and organises them into ten questions, ... that discuss what we know about mathematics teaching and learning around the world – and how these data might help you in your mathematics classes right now. The questions encompass teaching strategies, student learning strategies, curriculum coverage and various student characteristics, and how they are related to student achievement in mathematics and to each other. Each question is answered by the data and related analysis, and concludes with a section entitled "What can teachers do?" that provides concrete, evidence-based suggestions to help you develop your mathematics teaching practice

This document, OECD (2016b), is at: https://www.oecd-ilibrary.org/docserver/9789264265387-en.pdf?expires=1591023411&id=id&ac-cname=guest&checksum=8D0497280A63D4D60EBAC77DC0C3BD75

The OECD examines how certain teaching and learning strategies relate to student performance in mathematics. They used

> ... student background data to look at the relationships between other student characteristics, such as students' gender, socio-economic status, their attitudes toward mathematics and their career aspirations, to ascertain whether these characteristics might be related to teaching and learning strategies or performance.

We now consider the 10 questions (Q1–Q10) posed by OECD (2016b) as concerns that might be raised by teachers of mathematics, and highlight some of the key issues that emerge when addressing those questions.

Q1. How much should I direct student learning in my mathematics classes?

Bietenbeck (2014) provided evidence that a mix of 'traditional' and 'modern' teaching practices are optimal in the classroom for mathematics. The former can increase students' factual knowledge and their competency in solving routine problems, while the latter can foster reasoning skills supporting solutions to more complex problems. Teacher-led lessons might work well for simpler mathematical concepts, but other, different, student-oriented methods are better suited for more difficult concepts. Thus, for example, lessons that rely on a textbook for explaining concepts followed by completion of the text book exercises provide only a form of teacher-directed learning. It is better to augment such lessons with new activities that allow group working, using new tools such as technology to reinforce students' understanding of mathematical concepts.

Q2. Are some mathematics teaching methods more effective than others?

So-called 'cognitive action' strategies for teaching involve students summarising, questioning, using reflective thinking and predicting when attempting to solve problems. They make connections between mathematical facts, procedures and ideas that can enhance learning and achieve a deeper understanding of concepts (see, e.g., Burge, Lenkeit, & Sizmur (2015). For example, the increased demand for Date Science at all levels, including proposals for a school curriculum in the subject, provides a way for cognitive action to be practised through exposing students to real-world problems that could enhance their problem solving skills (see PISA in Jamaica and developing countries).

Q3. As a mathematics teacher, how important is the relationship I have with my students?

The findings from PISA show that classes with a better disciplinary climate produce better performance by students in tests in mathematics. Having a strong positive relationship with a class is clearly advantageous for the promotion of mathematics and getting students more familiar with the benefits from mathematics in later life.

Q4. What do we know about memorisation and learning mathematics?

PISA shows that students around the world often use memorisation to learn some aspects of mathematics. Students who are not naturally drawn to mathematics, or are not confident in their abilities in the subject, tend to use memorisation. Students with positive attitudes, motivation or interest in problem solving in mathematics or those who have little anxiety towards mathematics, are less likely to use memorisation techniques to try to master the subject. In general, boys are less likely than girls to use memorisation strategies. Memorisation seems to work well for the easier aspects of mathematics, but the evidence suggests that more challenging problems are answered less well by those students who use this strategy. Thus teachers should encourage students to go beyond rote memorisation and to think more deeply in order to make connections with real-world problems.

Q5. Can I help my students learn how to learn mathematics?

The evidence suggests that helping students learn how to learn helps them be successful in mathematics. Methods that encourage students to set their own goals and track their own progress help learners control their own learning. Research evidence shows that students are strategic in acquiring whatever superficial and deep understanding is needed to complete homework or pass exams (see, e.g., Hattie, 2009). In addition, when only surface-level mathematics is being tested, students are reluctant to venture into deeper-level learning without incentives. Thus, teachers should ensure that their teaching methods do not prevent students from adopting control strategies. In fact, teachers should ensure students reflect on how they learn and encourage them to discuss their problem solving procedures with their teacher and peers.

Q6. Should I encourage students to use their creativity in mathematics?

This is an area generally called 'elaboration strategies', encouraging students to link what they are learning in mathematics to their own experiences and real-world situations. Such strategies can be useful in helping students understand new information in mathematics and retain it over a long period. Elaboration can benefit students by helping them solve more difficult problems, but does not seem to have a positive effect on performance in solving easier problems. Thus, elaboration strategies should be encouraged when students are going to try to solve challenging problems.

Q7. Do students' backgrounds influence how they learn mathematics?

According to PISA, disadvantaged students are less frequently exposed to both applied and pure mathematics when compared with their more advantaged peers. For example, on average across OECD countries that took part, about 65% of socio-economically advantaged students, but only 43% of disadvantaged students, reported that they know well, or have often heard of, the concept of quadratic functions. Additional support for teachers in disadvantaged schools would be beneficial, and including ways to offer more individualised support to students who are struggling. Teachers also need more support to use pedagogies, such as flexible grouping, that increase the learning opportunities for all students.

Socio-economic background and status is associated with students' performance in PISA tests. Thus teachers should take into account socio-economic background of students and try to bridge the gaps in knowledge that these students often have compared to more advantaged students. At the same time, teachers should not shy away from more challenging mathematical topics, nor ignore the contextualised mathematical knowledge students bring from their everyday life and should make all students aware of the benefits of mathematical knowledge in later years of life.

Q8. Should my teaching emphasise mathematical concepts or how those concepts are applied in the real world?

This question relates to the continuous debate about the merits, or otherwise, between pure and applied mathematics and the extent to which different countries teach these two topics with a mathematics curriculum. The PISA data suggest that greater exposure to pure mathematics increases the chances that a student will be a top performer in the tests, and reduces the chances the student will be a low achiever. In some countries, greater exposure to applied mathematics was associated with poorer student performance in the tests. These results could, however, relate to the kinds of questions asked in the PISA tests.

In spite of this, applied mathematics can benefit and help enable students to solve problems that are multi-faceted because it can teach them how to question, make connections and predictions, conceptualise, and construct models to interpret and understand real situations. Thus, it is advisable for teachers to provide students with a variety of problems in context and pedagogies that involve project- or problem-based learning using real-world scenarios. We shall return to this in a later chapter where we discuss DS as a discipline that uses mathematics, but most importantly requires a different way of thinking. It is a discipline where its very existence is to solve real-world problems using data – numbers in context.

Q9. Should I be concerned about my students' attitudes towards mathematics?

According to PISA, mathematics is not often students' favourite subject. Under 40% of students said they studied mathematics because they enjoyed it and just over 40% believe they are not good at the subject. Negative attitudes towards mathematics lead to issues when students try to solve challenging problems. The OECD suggests that teachers might want to consider students' experience in problem solving in real-world contexts, and should explore innovative teaching tools for mathematics. It states that

Technology, including dynamic graphical, numerical and visual technology applications, can help students visualise mathematics problems while increasing their motivation or interest in the topic'. We agree with this and show in a later chapter that problems in Data Science provide a pedagogical context for students to learn problem solving using real data – often using data visualisation software which of itself can helps motivate the students.

Q10. What can teachers learn from PISA?

The OECD provides five recommendations (R1–R5 below) that teachers should try to put into practice. We list these verbatim.

R1: Make sure your teaching and assessments are balanced so that students can develop all the skills they will need for their future learning. Use multiple types of assessments, including oral tests, collaborative problem-solving and long-term projects, in addition to traditional written exams. Even standardised tests can be used occasionally to compare the performance of your class with students from other classes, schools, districts, cities and countries. Take advantage of questions from PISA that have been made public by the OECD or from PISA for Schools exams to serve this purpose.

R2: Yes, teachers are constrained by curricula and standards in a way that PISA is not, but it's still worth asking yourself: 'What is important for citizens to know and be able to do in situations that involve mathematics?' Even within the confines of your curriculum, this kind of thinking could help you decide which topics to present to your students – and how to present them – so that your students are better prepared for future learning and life outside of school. Many teachers around the world already emphasise abilities and skills in mathematics with the purpose of assessing both students' content knowledge (what they know) and applied knowledge (what they can do with what they know). Again, reading some assessment questions released by PISA might give you additional ideas for your class.

R3: Teach and assess students in ways that are fair and inclusive for everyone, regardless of gender, socio- economic background or ability. This guide provides several different recommendations related to fairness based on various student characteristics, but teaching in a more inclusive manner can be as simple as explaining content using different perspectives, evaluating students in different ways, and always taking into consideration students' background.

R4: Teachers do not teach in a vacuum. Every day you have classrooms of students who can inform your teaching, and colleagues down the hall or in the staff room who have different – and perhaps complementary – expertise and experience. In a supportive school environment, you are not solely responsible for your students' success. Listen to your students, collaborate with other teachers, participate in school decision-making, communicate with parents and learn from experts in your field.

R5: Don't let the constraints of a national curriculum or national exams limit your or your students' creativity. It is possible to innovate with tools and pedagogies. New approaches to teaching are tried and tested all the time, with varying degrees of success. If you are nervous, read up on strategies that have been successful for other teachers but might be new to you, or participate in innovation networks. Once you are more confident with the risks and rewards associated with innovation in teaching, you will be the one developing new strategies and resources for your colleagues to try.

How might Jamaica and other 'developing' countries fare if they enter PISA? It is tempting to try to predict how, for example, Jamaica might perform if it takes part in the mathematics test of PISA. The OECD has, in the past used per capita GDP as a predictor of mean performance. However, extensive research in this area shows that measuring outputs without accounting for inputs, using multilevel modelling, can produce untrustworthy results in educational league tables (Goldstein, 2004a,b).

Nonetheless, the OECD has modelled mean performance in science by using a nonlinear regression model with predictor variable of per capita GDP. Subject to the caveat that the model will be sub-optimal, Figure 2.3 shows the graph of mean performance in science plotted against per capita GDP for the 69 countries that took part in PISA in 2015.

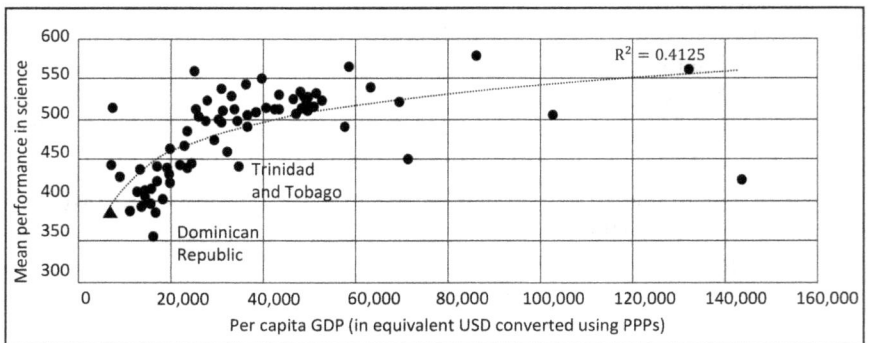

Figure 2.3 Mean performance in science and per capita GDP.

The dotted line represents the fitted model for mean performance regressed on per capita GDP. The line is not a particularly good fit with just over 41% of the variability in mean performance explained by the dotted line. However, from this model, the OECD predicts (see black triangle) that Jamaica's mean performance would be about 370. It would have been helpful if confidence intervals had been used to give an estimate of uncertainty in the prediction.

It should be borne in mind that there is a cost to participate in PISA. For example, in 2011–2013 England was paying £200,000 per year in contributions to the OECD to develop and run the international programme. In 2012, it cost England £575,000 to administer the test in the 170 schools who elected to take part, from the 192 selected by the OECD; more details are available in the document obtained under the Freedom of Information Act at: https://www.whatdotheyknow.com/request/pisa_cost_and_participation

However, the cost for Jamaica and other 'developing' countries to participate should be substantially lower than England's contribution in 2012, as they tend to have low GDP per capita.

Conclusions

The OECD spends a great deal of effort and money on promoting, running and reporting the PISA project and relies on a large number of countries to contribute financially. In addition, from each country's perspective there are large commitments in time and effort for schools, teachers and students to take part. For Jamaica, and other developing countries, they would need to investigate the cost of joining the programme and decide whether it was cost effective – and if the results could be used to enhance the education of the students and the pedagogy of teachers.

It is clear that teachers of *mathematics* in Jamaica and developing countries may be able to learn from the experience of teachers in other countries. By studying the PISA reports from 2000 to 2018, they could provide incentives for teachers and policy makers to benefit from joining the scheme.

The authors of PISA publish a number of supplementary reports and make a number of specific recommendations that, if followed, enable teachers to learn from the pedagogic experience of other teachers in a wide range of countries throughout the world.

There are statistical issues with the creation of the country mathematics scores and rankings based on the results of the PISA tests. These should be borne in mind when the results and consequent rankings of countries are produced. In particular, policy makers in a country should not get 'hung up' on the mathematics ranking position of that country because of the uncertainty attached to individual rankings.

References

Anderson, L. W., & Krathwohl, D. W. (2001). *A taxonomy for learning, teaching, and assessing: A revision of Bloom's educational objectives*, New York, Longman.

Bietenbeck, J. (2014). Teaching practices and cognitive skills. *Labour Economics, 30*, 143–153.

Bloom, B. S., Englehart, M., Furst, E., Hill, W., & Krathwohl, D. R. (1956). *Taxonomy of educational objectives: The classification of educational goals, by a committee of college and university examiners. Handbook I: Cognitive domain*, New York, Longmans, Green.

Burge, B., Lenkeit, J., & Sizmur, J. (2015). *PISA in practice—Cognitive activation in maths: How to use it in the classroom, national foundation for educational research (NFER)*, England and Wales, Slough.

Donoho, D. (2017). 50 years of data science. *Journal of Computational and Graphical Statistics, 26*(4), 745–766.

Goldstein, H., & Spiegelhalter, D. (1996). League tables and their limitations: Statistical issues in comparisons of institutional performance. With discussion. *Journal of Royal Statistical Society, A, 159*, 385–443.

Goldstein, H. (1997). Methods in school effectiveness research. *School Effectiveness and School Improvement, 8*, 369–395.

Goldstein, H. (2004a). International comparisons of student attainment: Some issues arising from the PISA study. *Assessment in Education: Principles, Policy and Practice, 11*(3), 319–330.

Goldstein, H. (2004b). Measuring educational standards. *Significance, 1*, 103–105.

Goldstein, H. (2019). PISA and the globalisation of education: A critical commentary on papers published in AIE special issue 4/2019. *Assessment in Education: Principles, Policy & Practice, 26*(6), 665–674. DOI: 10.1080/0969594X.2019.1674244.

Goldstein, H., Carpenter, J. R., & Browne, W. J. (2014). Fitting multilevel multivariate models with missing data in responses and covariates that may include interactions and non-linear terms. *Journal of the Royal Statistical Society: Series A (Statistics in Society), 177*(2), 553–564.

Hattie, J. (2009). *Visible learning: A synthesis of over 800 meta-analyses relating to achievement*, London and New York, Routledge.

Marriott, J. M., Davies, N., & Gibson, L. (2009). Teaching, Learning and Assessing Statistical Problem Solving. Online at https://ww2.amstat.org/publications/jse/v17n1/marriott.html

OECD. (2016a). How teachers teach and students learn: Successful strategies for school. OECD Education Working Papers, No. 130, Paris, OECD Publishing; https://www.oecd-ilibrary.org/education/how-teachers-teach-and-students-learn_5jm29kpt0xxx-en

OECD. (2016b). *Ten questions for mathematics teachers … and how PISA can help answer them*, Paris, OECD Publishing; https://dx.doi.or/10.1787/9789264265387-en

Hopfenbeck, T. N., Lenkeit, J., El Masri, Y., Cantrell, K., Ryan, J., & Baird, J.-A. (2018). Lessons learned from PISA: A systematic review of Peer reviewed articles on the programme for International student assessment. *Scandinavian Journal of Educational Research, 62*(3), 333–353.

Spiegelhalter, D. (2013). https://www.statslife.org.uk/features/1074-the-problems-with-pisa-statistical-methods

Chapter 3

Finland: policies for education and mathematics teaching and innovation

Leo Pahkin

Setting the scene

At the core of developing successful education systems, which promote sustainable economic growth, are equitable structures that allow all students to succeed. Many countries have high academic achievement levels for specific groups of students but it is less common for countries to have homogenous results for students of different areas (e.g., rural/urban), gender, and socio-economic home backgrounds. This chapter will provide an outline of the main characteristics of mathematics education in the Finnish system to examine the key factors that have led to equitable mathematics results for students. This will include consideration of school organisation, teaching and learning strategies for mathematics and teacher training.

Educational policy in Finland

At the centre of educational policy in Finland is a focus on equality (Malaty, 2006). Despite a lack of emphasis on high performance in international testing, Finland has been highlighted as having a very successful education system from pre-school through to Higher Education.

A key reason for the high regard of the Finnish educational system is the recent positive performance in the Programme for International Student Assessment (PISA) tests in Mathematics (and Science and Native Language). PISA measures mathematical literacy (see Chapter 2) and a particular characteristic of Finland's results are a low spread of student achievement with only a very small percentage of students demonstrating low achievement (Malaty, 2006).

Reforms made in the Finnish educational system have occurred over the past 50 years from when the country was in a weak financial position and was characterised as a 'developing economy'. Finland underwent a serious economic crisis with the fall of the Soviet Union, which influenced the educational system, but a national policy from the 1990s emphasised both mathematics and science as priorities.

A key vision for Finland was a high-tech economy that was influenced by the rise of Nokia (Hannula, Lepik, Pipere, & Tuohilampi, 2013). The decision though was taken to invest in the education of all the people in Finland, whatever stage of life they are in and this overriding philosophy has never changed and is an important characteristic aspect of Finnish people and the education system.

The complete educational system in Finland is shown in Figure 3.1 (Education in Finland, 2017).

In this chapter, we will focus on the Basic Education stage (age 7–16). However, similar principles to those described in this chapter apply throughout all levels of education. The next section will provide an overview of the schooling systems drawn from *This is Finland* (2016). Compulsory schooling starts from the school year during which children turn age seven and ends when they have either completed the entire nine-year basic education curriculum or, at the latest, after the school year when they turn age 17. All children are also entitled to one year of pre-school education. Tuition, schoolbooks, learning materials and equipment are all provided free of charge for the 9 years of basic education. All pupils are also provided with a free school meal every school day. School days must not be longer than five lessons for children in first- and second-grade classes, and seven lessons for older schoolchildren. Each lesson is 45 minutes. There are no nationwide examinations or grading tests. There are a total of 190 school days in a Finnish school year. The school year starts in the middle of August and ends in May. Finnish children have about 10 weeks of summer holiday as well as holidays in autumn, Christmas break and winter, usually in February. In Finland, almost all youngsters complete the syllabus of basic education and graduate from comprehensive school.

A career pathway for teaching is popular and high standards are set for entrance. For example, in 2014 only 9% of applicants sitting the entrance exam for Helsinki University's training for class teachers were admitted. Teachers working with children in grades 1–6 must be qualified to a level of at least Master of Education. Teachers working with grades 7–9 must have a Master's degree in their subject, as well as high-level qualifications in education.

In contrast to other countries, there are a number of unusual policies relating to general education in Finland (see Sahlberg, 2016, for more details). The following section will identify and expand on these policies in more detail.

Unique Finnish educational policies

No standardised testing

Finland has almost no standardised tests. The only exception is the National Matriculation Exam, which has one compulsory test in mother tongue and literature and at least four free choice tests in other subject for learners at the end

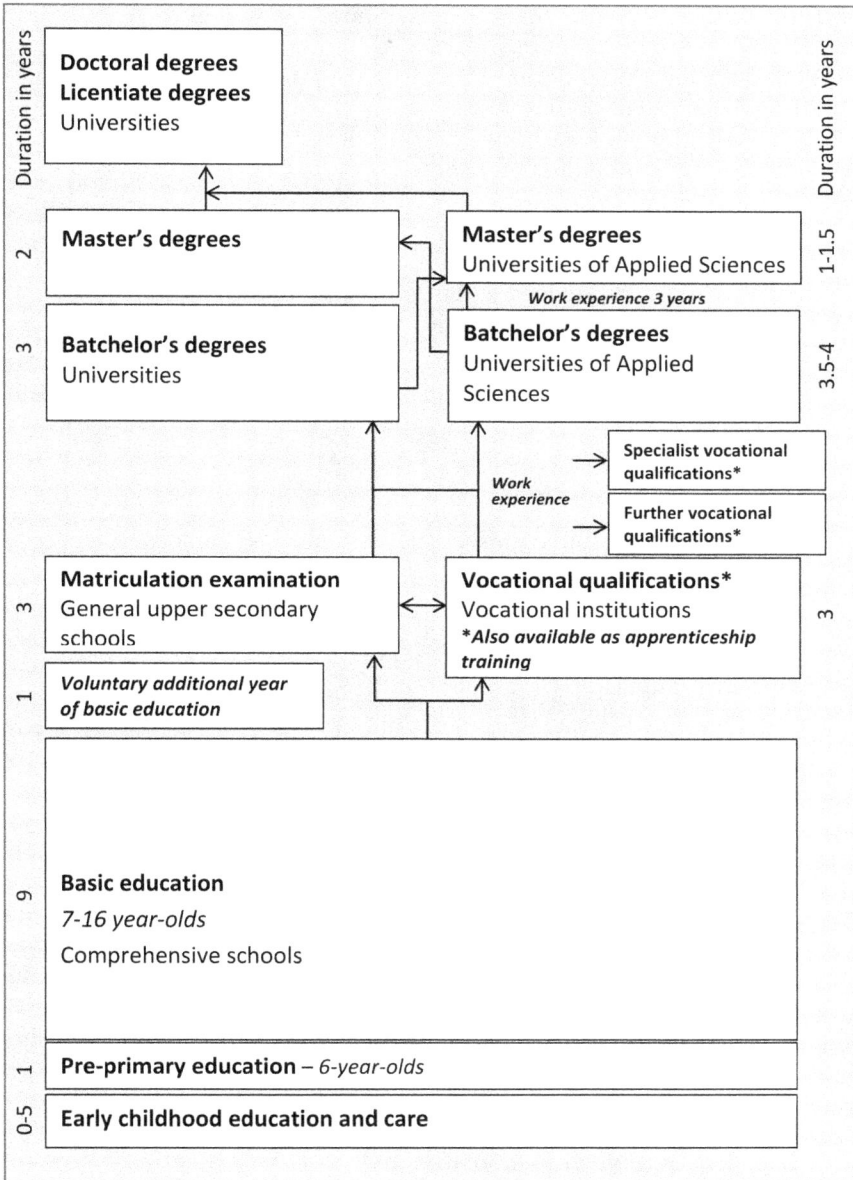

Figure 3.1 Educational system in Finland (duration in years shown).

of their time at upper-secondary general school (equivalent to an American high school). Around half of the relevant age group chooses upper-secondary general education. All children throughout Finland are graded on an individualised basis and a grading system set by their teacher. Tracking overall progress is done by the evaluation centre of education (Karvi), which samples groups across different ranges of schools.

Accountability for teachers

All teachers are required to have a Master's degree before entering the profession. Teaching programmes are the most rigorous and selective professional training courses in the entire country. If a teacher in a school is not performing well, it is the individual principal's responsibility to do something about it.

Cooperation not competition

While many Western countries see competition as an important aspect of education policy with, in particular, competition between schools at secondary level to attract more and talented learners, the Finns see it differently with schools working cooperatively and collaboratively together and learners just going to their local school. This attitude is one of the key factors that has placed Finland at or near the top in PISA rankings. Finland's educational system is not concerned about artificial or arbitrary merit-based systems. There are no lists of top performing schools or teachers. It is not an environment of competition – instead, cooperation is the norm.

A focus on equality

Many years ago, the Finnish school system was in need of some serious reforms. The programme that Finland put together focused on returning to the basics. It was not about dominating with excellent marks or 'upping the ante'. Instead, educators looked to make the school environment a more equitable place. Since the 1980s, Finnish educators have focused on prioritising these basics:

- Education should be an instrument to balance out social inequality;
- All students receive free school meals;
- Ease of access to healthcare;
- Psychological counselling;
- Individualised guidance.

Beginning with the individual in a collective environment of equality is Finland's way.

Providing professional options past a traditional college degree

Finland solves the dilemma of academic or vocational training at age 16+ by offering options that are equally advantageous for each learner continuing their education. There is less focus on the dichotomy of college-educated versus trade-school or working class. Both can be equally professional and fulfilling for a career.

In Finland, there is the Upper Secondary School which is a 3-year programme that prepares students for the Matriculation Test that determines their acceptance into a University. This is usually based on specialties they have acquired during their time in 'high-school'. Next, there is vocational education, which is a 3-year program that trains learners for various careers. They have the option to study also in upper-secondary general school at the same time and then take the Matriculation test. There is also the possibility of applying to University or Applied University after graduation.

Shorter school days

Learners in Finland usually start school anywhere from 0900 to 0945 hours. Research has shown that early start times are detrimental to students' well-being, health and maturation. Finnish schools start the day later than those in many other countries and usually end by 1400–1445 hours. They have longer class periods and much longer breaks in between. The overall system is not there to ram and cram information to their learners, but to create an environment of holistic learning.

Consistent instruction from the same teachers

Learners in Finland often have the same teacher for up to **6 years** of their education. During this time, the teacher can take on the role of a mentor or almost a family member. During those years, mutual trust and bonding are built so that both parties know and respect each other.

Different needs and learning styles vary on an individual basis. Finnish teachers can account for this because they are aware of the learner's own idiosyncratic needs. They can accurately chart and care for their progress and help them reach their goals. There is no passing along to the next teacher because there is no 'next teacher'.

A more relaxed atmosphere

There is a general trend in what Finland is doing with its schools towards less stress, less unneeded regimentation and more caring. Learners have several times to eat their food, enjoy recreational activities and generally just relax. Spread throughout the day are 15- to 20-minute intervals when they can get up and stretch, get some fresh air and decompress.

This type of environment is also needed by the teachers. Teacher rooms are set up all over Finnish schools, where they can be comfortable and can relax, prepare for the day or just simply socialise. Teachers are people too and need to be functional so they can operate at the best of their abilities.

Less homework and outside work required

According to the OECD (Organisation for Economic Co-operation and Development) learners in Finland have the least amount of outside work and homework of any other learners in the world. They spend only half an hour a night on work for school. Finnish learners are getting everything they need to get done in school without the added pressures that come with excelling at a subject. Without having to worry about grades and busy-work they are able to focus on the true task at hand – learning and growing as a human being.

Teacher training

An important component of any education system is the teacher training system. The current structure for Finnish teacher education is summarised below (more details are given by Beatty and Ferreras (2018) when they compare and contrast teacher training in Finland and the United States):

- All teachers take a 5-year course of training to become a qualified teacher;
- Primary teachers take a 180 credits at Bachelor's level and 120 credits at Master's level focused on Pedagogical studies (but with mathematical studies);
- Secondary teachers take a 180 credits at Bachelor's level in their major and minor specialist subjects (e.g., Mathematics and Physics) and 120 credits at Master's level focused on subject pedagogy.

Trainee teachers also have guided teacher training periods in school, which accounts for 20 credits of their 300-credit Master's Degree. Although this is a relatively small component, it is both meaningful and important with a stress on combining theory and practice.

There are nine universities located around the country that provide teacher education. Training periods are in the teacher training schools. These are state schools with extra capacity for having trainee teachers and expert teacher mentors to oversee the school-based work. These schools are similar in conception to University Hospitals and sometimes referred to as University Practice Schools, with the emphasis on the trainees working alongside expert teachers with time for trainees to reflect on their own development as a teacher.

Normally teaching practice takes place in the first, third and fifth years of the Masters programme with the first- and third-year practice taking place in the University teacher training schools under the guidance of professional mentors, whilst the final practice takes place in a network of field schools associated with the University. Trainee teachers work in pairs through most of these experiences, learning from each other and from both their University and school-based supervisors.

It is important to note that research is highly valued in Finland's education system and emphasised both in training and in schools. Trainee teachers not only learn about the research literature in their areas of study but, as part of their Master's degree, they have training in formal research skills. Teachers are also encouraged to undertake doctoral degrees as part of their normal classroom work and have freedom to experiment and evaluate their new approaches to teaching and learning.

Finally, it should be stressed that competition for places in teacher training is strong, particularly at Primary level where only about 10% of applicants gain places. Education is regarded in Finland as a very highly respected profession on a par with doctors and lawyers and this is reflected in the quality of the students entering the profession.

Encouraging mathematical thinking

By now, you will have a clearer idea of background to the education aims and provision in Finland, so it is time to look more closely at how this translates into mathematics teaching and learning. We start though with points on 'Strategies for Teachers to encourage Mathematical Thinking' from the Finnish National Board for Education (2016), which provides the underpinning philosophy for all our mathematics provision:

- If you want to increase curiosity, *allow questioning;*
- If you want to develop problem-solving skills, *link school knowledge to real-life problems and encourage pupils to work together to seek solutions;*
- If you want to increase understanding, *combine knowledge and skills from different subjects;*
- If you want to raise citizens who will develop society, *promote inclusiveness and participation, give opportunities to make a difference, and facilitate positive – not negative – critical thinking;*
- If you want to strengthen learners' self-confidence and learning motivation, *give constructive and honest feedback. Never humiliate or put down a learner.*

Note also though that teachers are encouraged to let their learners, whatever their grade and ability, explore ideas for themselves, along the lines of:

- Start with a problem (instead of theory);
- Set an interesting and reachable target;

- Activate students;
- Choose solvable problems, but not ordinary ones;
- Use a constructive set of problems with higher level problems connected to the earlier problem;
- Use project working and technology;
- Let the learners explore!

Not all teaching will follow this model but we are always emphasising that the learners should explore, learn, consolidate and extend rather than being taught!

The following problems illustrate this approach with the underlying aim being to help our learners to become mathematical thinkers:

Grades 1–2

Here we start to emphasise mathematical thinking through problems in logic.

Example: What are the missing shapes in these diagrams (Figure 3.2)? Justify your answers.

Grades 3–6

Here emphasis is put on numerical investigations to enhance mathematical and logical thinking as shown in the following example:

Example: How many numbers can you find such that:
- It has three digits.
- It only uses the digits 0, 1, 2, 3, 4 and 5.
- The second digit is divisible by 2.
- The hundreds and units digit are the same.

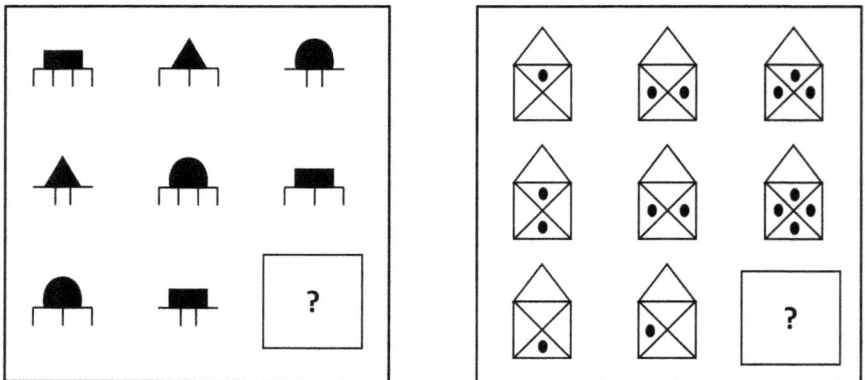

Figure 3.2 Logical problems for learners to discuss.

- Having found these numbers, calculate the sum of the digits.
- Which one of these statements is **TRUE**?
  - A: the sum of the digits is always **EVEN**
  - B: the sum of the digits is always **ODD**
  - C: the sum of the digits can be **EVEN or ODD**

Grades 7–9

Now a more practical context is used to encourage mathematical thinking such as:

Example: CASINO

I have two fair dice. You pay €1 to throw the two dice.
I will pay you:
- €2 if you get two numbers the same;
- €1 if you get a sum total more than 8.

Do you want to play?

Here the aim is to move towards 'expected value' and the class will play this in pairs with everyone creating their own game and analysing the results.

We also encourage all our learners not to be afraid of multi-stage word problems and help them with strategies to help solve the problem such as that given below.

Procedure for solving word problems

1 Figure out what is given in the problem and what is asked for.
2 Draw a picture or other illustration (or use manipulatives).
3 Do all the necessary steps for the solving process.
4 Find the solution(s) and the reasons behind it.

The teacher should make the problem-solving visible so that it becomes easier for all the learners to understand the process, to learn to communicate clearly and concisely and for them to be aware of the process of problem solving and reflect on it.

Here is an example to demonstrate this process with an emphasis on the local context:

Example: SKI LIFT

A total of 20 skiers are to get to the mountain using eight-bench ski lifts. Each bench can fit four skiers at the most but each bench must have at least two skiers.

Find out how all the skiers can travel to the mountain.

Solve the problem and reflect on the problem-solving process and the strategies you used in finding the solution.

Illustrate and communicate your problem-solving process as accurately as possible.

The aforementioned examples give you some idea of how mathematics is taught and learnt in Finnish schools. It should be noted though that our teachers have autonomy and are trusted and valued. Teachers have freedom to try out and evaluate different approaches and many will be undertaking research in effective teaching and learning as part of their work for a higher degree.

Above all though, teachers ensure that all their learners are engaged in the work and any learners needing more help are identified and, for example, given more support or revision work. All schools have a duty to encourage and help every student and intervene with extra support when needed, working collaboratively with parents and carers where appropriate. The underlying theme is to ensure no learner is left behind.

This means that schools have a policy of ensuring that teachers are aware of the progress of all their learners and have a tiered policy of support, including providing extra support for any slow learner in the classroom, supporting the learner outside normal lesson time and discussing the progress of learners with their parents to see what more can be done to enhance their learning.

Ongoing reforms and changes

Despite Finland's success in international comparisons, we embarked on a new wave of reforms to build on the strengths of the education system and, at the same time, meet the challenges of a rapidly changing and complex world. The reforms were developed with emphasis on the meaningfulness of learning, the engagement and well-being of learners as well as educational equality.

From the learners' perspective, the focus has been on ways to improve the motivation and relevance of learning, enhanced thinking skills as well as other transferrable skills (Figure 3.3). We also wanted schools to develop as collaborative learning communities and this is reflected in an integrated multi-disciplinary pedagogical approach with an emphasis on crossing the boundaries of subjects. This is illustrated in Figure 3.3.

It has been a challenge to encourage students' active participation in the everyday life of their school and to change the traditional roles of teachers and students to create a school culture where students have enough room to explore, think, collaborate and create. More detailed information is given by Halinen (2018).

Although it might seem strange to the outsider that we continue to make significant reforms to an internationally praised educational system, we seek to build on our strengths and move forward with our culture of trust, support and collaboration central to our reforms.

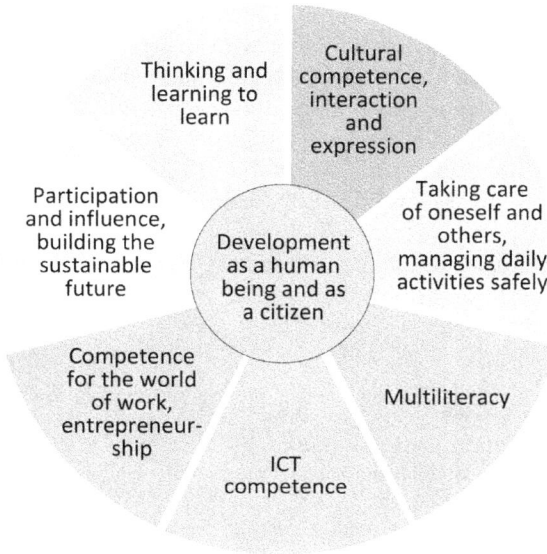

The diagram shows a wheel with "Development as a human being and as a citizen" at the center, surrounded by segments:
- Thinking and learning to learn
- Cultural competence, interaction and expression
- Taking care of oneself and others, managing daily activities safely
- Multiliteracy
- ICT competence
- Competence for the world of work, entrepreneurship
- Participation and influence, building the sustainable future

Figure 3.3 Transferrable skills.

Summary and conclusions

We have presented here an overview of the Finnish Education system and in particular, the underlying themes and philosophy for the teaching and learning of mathematics. There is no doubt that we have an Education system that works well but it has taken several decades to evolve and, compared with many other countries, it is expensive. Only 50 years ago, we would have been classed as a developing country and there was, even at that time, public support for investing in education so that the next generations of young people would have the skills and knowledge to help the country to become more developed and financially secure.

This has worked well and, although the move to having little of no competition between schools, so effectively no 'private' schools, took some time to implement, we are now fortunate that we have an educational system to be proud of, with its success based on the availability of this high standard of education to children throughout the country, regardless of geographic location or socio-economic background.

As pointed out above, the potential downside has been the financial cost of education but the population is overwhelmingly in favour of prioritising education as a key policy for the country without concern for the cost. We have performed well in international tests in Mathematics, Science and Native Language but this was not our aim but rather an interesting by-product of the educational system that we have adopted.

Our aim has, and will continue to be, to ensure that all learners are well supported and given equal opportunities to make progress in all aspects of education, with mathematics being a key component. It is also worth noting that the teaching profession is held in high regard, akin to high professionals, and this is important for the sustainability of the success in our educational provision for all learners.

In the words of our Minister of Education, Andersson (2020),

> 'everyone in Finland can trust that their own local school is among the best in the world'.

References

Andersson, L., Interview for Neues Deutschland. (2020). https://www.transform-network. net/en/blog/article/principled-pragmatism/

Beatty, A. & Ferreras, A. (2018). Supporting Mathematics Teachers in the United States and Finland; https://www.nap.edu/read/24904/chapter/5

Education in Finland: Key to the nation's success. (2017). https://toolbox.finland.fi/life-society/finfo-education-finland-key-nations-success/ and https://www.oph.fi/sites/default/files/documents/finnish_education_in_a_nutshell.pdf

Finnish National Board for Education. (2016). Curriculum Reform in Finland; http://www.euroedizioni.it/attachments/article/697798/Curriculum%20Reform%20in%20Finland.pdf

Halinen, I. (2018). The New educational curriculum in Finland. In: *Improving the quality of childhood in Europe* (Vol. 7). Brussels, Alliance for Childhood European Network Foundation; https://www.oph.fi/en

Hannula, M.S., Lepik, M., Pipere, A. & Tuohilampi, L (2014). Mathematics Teachers' Beliefs in Estonia, Latvia and Finland. In *CERME 8, 6-10- February 2013, Manavgat-Side, Antalya - Turkey: WG (Working groups) 11 Papers*. ERME, CERME 8 Eighth Congress of European Research in Mathematics Education, Manavgat-Side, Antalya, Turkey, Turkey, 06/02/2013. http://cerme8.metu.edu.tr/wgpapers/WG11/WG11_Hannula.pdf

Malaty, G. (2006). *PISA results and school mathematics in Finland: Strengths, weaknesses and future.* Joensuu, University of Joensuu.

This is Finland: The truth about Finnish Schools. (2016). https://finland.fi/life-society/the-truth-about-finnish-schools/

Sahlberg, P. (2016). *Finnish lessons 2.0: What can the world learn from educational change in Finland?* New York, Teachers College Press, Columbia University. ISBN **9780807755853**

Chapter 4

Problem-solving approach and lesson study for improving performance in mathematics

Masataka Koyama

Introduction

Proficiency in mathematics is an important requirement to capitalize on opportunities for sustainable economic growth and job creation in the world. The quality of education cannot exceed the quality of teachers.

This chapter will share the methods of the problem-solving approach and lesson study used in Japan for developing students' mathematics thinking and their ability for mathematical problem solving and for promoting professional development of mathematics teachers and educators.

Background

The school education system in Japan comprises 6 years of education at primary school, 3 years at lower secondary school and 3 years at upper-secondary school. The 9 years spent at primary and lower-secondary schools, for students aged from 6 years to 15 years, are compulsory education for all children. The national curriculum standard is prescribed in the Course of Study determined and issued by the Ministry of Education.

The Course of Study for Mathematics (MCS) has been revised and reissued approximately once every 10 years since the establishment of the Constitution of Japan and the Fundamental Law of Education in 1947 (e.g., Ministry of Education, 2008a, 2008b, 2009).

Figure 4.1 shows one cycle of revising the MCS. The shaded area in the figure indicate some agents involved in the process of revision cycle.

School textbooks must be approved by the Ministry of Education in line with the Course of Study. Public school teachers are local prefectural or municipal officials and are appointed by the respective local prefectural or municipal boards of education. Primary school teachers teach almost all school subjects at their own grade while secondary school teachers teach their major school subjects.

In the following sections, we look at the problem-solving model for the teaching and learning of school mathematics (Koyama, 2008a, 2010) and

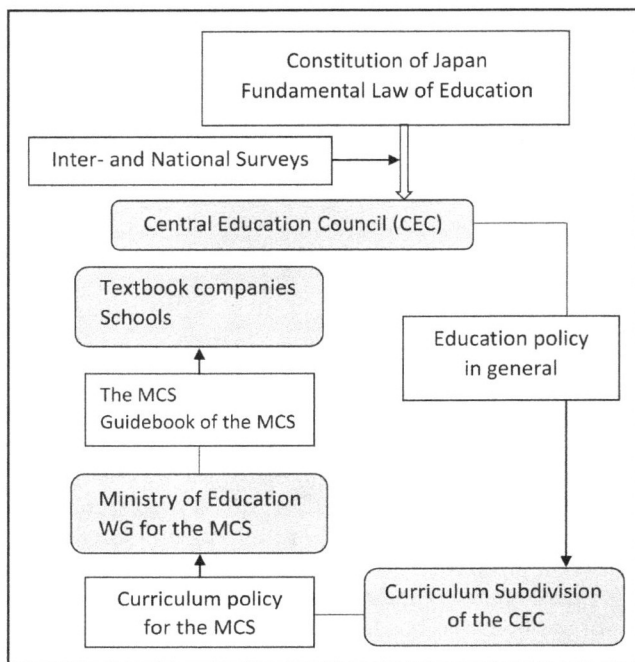

```
┌─────────────────────────────────────────────────────────────┐
│                    ┌──────────────────────────┐              │
│                    │   Constitution of Japan   │              │
│                    │ Fundamental Law of Education│            │
│                    └──────────────────────────┘              │
│  ┌──────────────────────────┐                                │
│  │ Inter- and National Surveys │───────►                     │
│  └──────────────────────────┘                                │
│        ┌──────────────────────────────────┐                  │
│        │  Central Education Council (CEC)   │                 │
│        └──────────────────────────────────┘                  │
│     ┌──────────────────────┐                                 │
│     │ Textbook companies   │         ┌──────────────────┐     │
│     │ Schools              │         │ Education policy  │     │
│     └──────────────────────┘         │ in general        │     │
│     ┌──────────────────────┐         └──────────────────┘     │
│     │ The MCS              │                                  │
│     │ Guidebook of the MCS │                                  │
│     └──────────────────────┘                                  │
│     ┌──────────────────────┐                                 │
│     │ Ministry of Education│                                  │
│     │ WG for the MCS       │                                  │
│     └──────────────────────┘                                  │
│     ┌──────────────────────┐     ┌──────────────────────┐    │
│     │ Curriculum policy    │     │ Curriculum Subdivision│    │
│     │ for the MCS          │     │ of the CEC            │    │
│     └──────────────────────┘     └──────────────────────┘    │
└─────────────────────────────────────────────────────────────┘
```

Figure 4.1   One cycle of revising the Course of Study for Mathematics (MCS) in Japan.

clarify and reflect on using lesson study to enhance problem-solving lessons in primary mathematics in Japan (Koyama, 2019).

Problem-solving approach for improving students' performance in mathematics

A typical model of good practice recognized by many Japanese educators and teachers is the problem-solving lesson in the mathematics classroom (Becker & Shimada, 1997; Stigler & Hiebert, 1999; Burghes & Robinson, 2009; Koyama, 2012). The problem-solving lesson has four distinct phases: presentation of the problem, development of a solution, progression through discussion and summarizing the lesson (Burghes & Robinson, 2009, p. 56). This might be characterized as the parallel and collaborative version of four steps identified by Polya (1945) in solving a mathematical problem: under-standing the problem, devising a plan, carrying out the plan and looking back and extending.

Why do many Japanese educators and teachers recognize the problem-solving lesson as a good model for the teaching and learning of mathematics? Sawada (1997) pointed out the advantage of using the problem-solving lesson model for teaching and learning mathematics in the classroom as follows:

> 'Students participate more actively in the lesson and express their different ideas or solutions more frequently. Students have more opportunities to make comprehensive use of their knowledge, skills, and ways of thinking. Even low achieving students can respond to the problem in some significant ways of their own. Students are intrinsically motivated to give their justifications or proofs. Students have rich experience in the pleasure of mathematical activities and receive the approval from peer students in the classroom (Sawada, 1997, p. 23)'.

The essence of doing mathematics is the process of solving a problem mathematically rather than a product. If acquired in the process of solving problems mathematically, then we believe that the mathematical knowledge, skills and ways of mathematical thinking are applicable in a new or unfamiliar situation for learners. In the mathematics classroom, therefore, a mathematics lesson as an integration of the teacher's teaching activity and the students' learning activity should be structured for the overall process of solving problems mathematically. The problem-solving lesson is a student-centred model of teaching and learning mathematics that may encourage students to construct mathematics collaboratively in a mathematics classroom using their naïve conceptions as well as their acquired mathematical knowledge, skills and ways of mathematical thinking.

Lesson study for the professional development of teachers as a life-long process

The lesson study of school mathematics has been recognized as an important cultural and collaborative means of the continuous professional development of teachers as a life-long process in Japan (National Association for the Study of Educational Methods, 2011). The Japanese lesson study model is well known internationally (Stigler & Hiebert, 1999; Lewis, 2002; Isoda, Stephens, Ohara, & Miyakawa, 2007; Shimizu, 2010; Takahashi, 2010, 2014). Figure 4.2 shows Japanese lesson study model (one cycle).

During lesson study, teachers collaborate to:

1 formulate long-term goals for student learning and development;
2 plan and conduct lessons based on research and observation to apply these long-term goals to actual classroom practices for particular academic contents;
3 carefully observe the levels of students' learning, their engagement and their behaviours during the lesson;
4 hold post-lesson discussion with their collaborative groups to discuss and revise the lesson accordingly (Lewis, 2002).

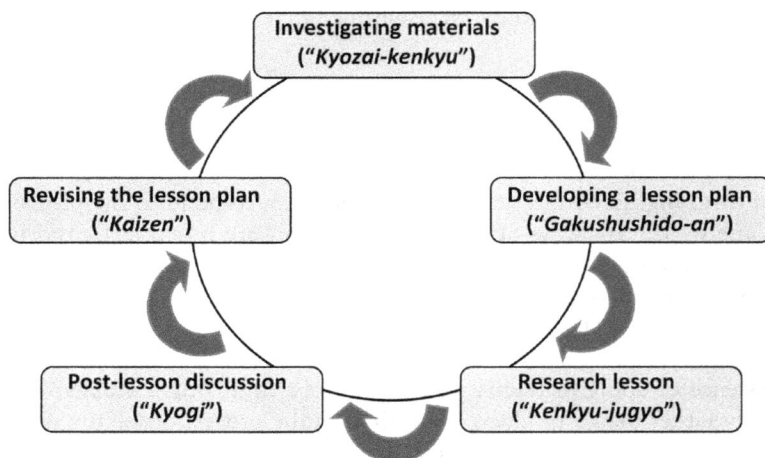

Figure 4.2 Japanese lesson study model (one cycle).

The success of lesson study primarily depends on improvements in teacher practice and the promotion of collaboration among teachers. Lesson study provides Japanese teachers with opportunities to make sense of educational ideas within their practice, to change their perspectives about teaching and learning and to learn to see their practice from the students' perspective (Takahashi, 2010). The lesson study can link together and promote both the pre-service teacher training at universities and the in-service teacher training at school, local district and nationwide levels (Koyama, 2008b; Corey, Peterson, Lewis, & Bukarau, 2010). The lesson study as in-service teacher training takes different forms.

In the following, as a mathematics educator who has been actively involved in various lesson studies in Japan, the author shares his experience of a school-based and cross-district lesson study (Koyama, 2015) of problem-solving lesson as part of primary school mathematics. The author then reflects on the overall process of lesson study by focusing on the relationship between collaboration and reflection.

An example of the lesson study of primary school mathematics

As one example, this section looks at a school-based and cross-district lesson study of primary school mathematics at the Hiroshima University Attached Primary School (HUAPS), where the author as a colleague has been collaboratively working with all the mathematics teachers of the school for several decades since the 1990s (Koyama, 2015).

The mathematics lessons at the HUAPS have been structured using the problem-solving lesson model for the teaching and learning of primary school mathematics in emphasizing students' continuous awareness of learning mathematics.

Characteristics of the HUAPS

The HUAPS is located in Hiroshima City in Hiroshima Prefecture. The school has been attached to Hiroshima University for more than 100 years and is tasked with taking part in practice teaching for pre-service teacher training at Hiroshima University, performing new developmental research for next-generation education, and conducting lesson study and opening classes to teachers from all districts in Japan. The school is a unique national primary school because it has three mathematics teachers who teach only mathematics at different grades, while other primary school teachers teach almost all school subjects at their own grade at local public schools. For more than 100 years, the school has issued its own monthly journal *School Education*. Each edition includes a pair of articles on the lesson study of primary school subjects written by one HUAPS teacher who conducted a research lesson and one Hiroshima University educator who is a specialist in school subject education. In addition, for a periodic seminar on primary school mathematics, we have organized a collaborative study team comprising not only mathematics teachers and mathematics educators but also graduate school students attending PhD and Master's courses on mathematics education at Hiroshima University.

Overall process of the lesson study at the HUAPS

Figure 4.3 shows one example of lesson study conducted by the collaborative study team at the HUAPS from November 2012 to February 2013 in the third semester of the 2012 academic year. The lesson adopted the problem-solving lesson model in the teaching and learning of the topic "Triangle" for second graders at the school. The overall process of the lesson study at the HUAPS comprised seminars before the research lesson, planning the research lesson, conducting the research lesson and post-lesson discussion and reflection on the research lesson.

The research lesson as a part of the lesson study at the HUAPS was conducted by a male teacher with 18 years of teaching experience and a Master's degree in education. After the four seminar sessions held by the collaborative study team, he designed a series of 15 lessons for the teaching unit "Triangle and Quadrangle" for his second graders of the school. He then selected the fifth of the 15 lessons as the research lesson and developed a lesson plan for the research lesson. The objectives of the lesson were to foster students' ability to think logically, especially through

| Seminar before Research Lesson | ...... | Collaborative Study Team for Primary School Mathematics in HU (Collaboration among mathematics teachers, mathematics educators, and graduate school students, HU) Teaching materials for geometry education in primary school |

↓

| Planning Research Lesson | ...... | Mathematics Teacher Mr. Kazushige Maeda Planning teaching unit of 'triangle and quadrangle' in 2nd Grade Making lesson plan for research lesson of 'triangle' in 2nd Grade |

↓

| Conducting Research Lesson | ...... | Mathematics Teacher Mr. Kazushige Maeda Conducting research lesson open to teachers from all districts Mathematics Educator Prof. Masataka Koyama and HU students Recording and making protocol of the research lesson |

↓

| Post-Lesson Discussion | ...... | Classroom teacher and teachers from other schools Discussion on the research lesson • Explanation of the aims of the research lesson • Question and answer among participants |

↓

| Reflection on Lesson Study | ...... | Articles for *School Education Journal*, HUAPS Reflection on the lesson study by Mr. Kazushige Maeda Reflection on the lesson study by Prof. Masataka Koyama |

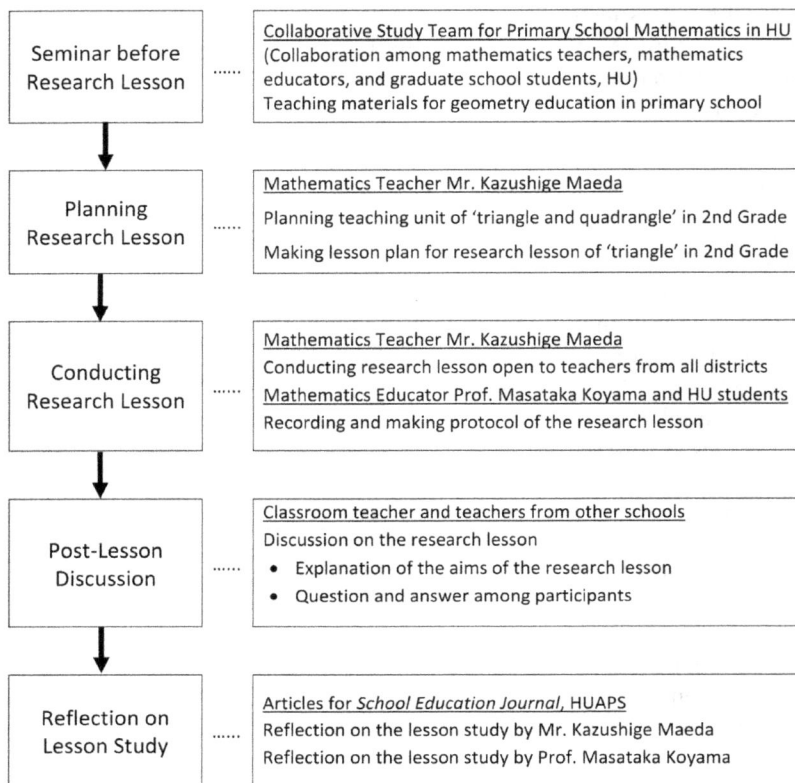

Figure 4.3 Overall process of lesson study at the HUAPS.

activity that stimulates the students to think about the definition of a triangle. The features of the lesson were to use different figures as teaching materials, to show those figures one by one and thus stimulate students to remember the definition of a triangle and to use explicitly the definition in explaining the reason for their judgment, and to incorporate the important question "Why would a figure that has three connected points not be a triangle?" to question the students' recognition of a triangle and deepen their understanding of the definition of a triangle.

Reflection on the lesson study at the HUAPS

In the following, we reflect on the details of the lesson study from three viewpoints: the collaborative study of teaching materials for the research lesson, the development of a lesson plan for the research lesson and the reflective analysis of the research lesson.

Collaborative study of teaching materials for the research lesson

In the first session of the seminar, the mathematics teacher proposed the students' activity to identify whether a figure is a triangle so that his students capture the concept of a triangle in terms of its components and shared with the team members his expectations to be realized through this activity. The main issue of the team discussion was what criteria should students use in judging whether a figure is a triangle. Finally, they recognized the necessity to investigate the mathematical background related to the definition of a triangle and the research findings on the merits and demerits of using counter-examples for the formation of the students' concept of a triangle.

In the second session, as a result of the team discussion, it was confirmed that the activity proposed by the mathematics teacher in the previous session could help the students learn a fundamental mathematical activity of going back to the definition if needed. Through the activity, the students might be encouraged to think about the reason why a triangle is not defined as a figure with three connected points. The team members then discussed whether an alternative teacher's activity of asking the students "Can you see a triangle in the given figure?" would be more appropriate for the students than the asking them "Is the given figure a triangle?" At the end, they recognized the necessity to design the development of a research lesson in more detail.

In the third session, it was confirmed that the planned research lesson could be theoretically supported by both the philosophy of mathematics education and the cognition theory of geometric figures in psychological research. The team members agreed that the research lesson could contribute to deepening the students' recognition and understanding of a triangle. They then discussed how the students would interpret the teacher's directive "Let's classify the given figures into two groups." Finally, the mathematics teacher proposed six different figures to be used in the research lesson (see Figure 4.4). The team members discussed an effective way of showing the figures to the students in the research lesson.

In the fourth session, the main issue of the team discussion was how to deepen students' understanding of the definition of a triangle such that a geometric figure surrounded by three straight lines is named a triangle which

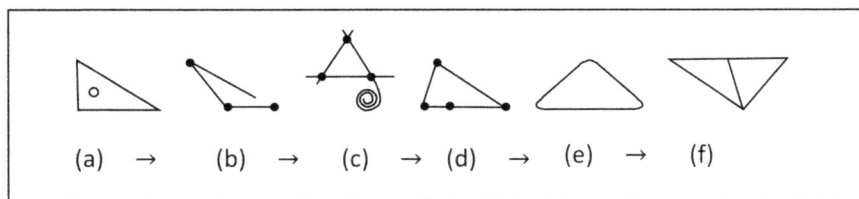

Figure 4.4 Intended order of showing the figures to the students in the research lesson.

was already learnt in the previous lesson in the second grade. In the discussion, the team focused on how to organize both the teacher's activity of shaping students' recognition and understanding of a triangle and the students' activity of rethinking the definition of a triangle. To promote these activities, they discussed a possible way of showing the prepared six figures and the expected merits and demerits of giving the students counter-examples of a triangle in the research lesson.

Development of a lesson plan for the research lesson

After the four sessions, the mathematics teacher of the research lesson developed the lesson plan. We must remember that before this research lesson, the second graders had already learnt the definition of a triangle in the teaching unit "Triangle and Quadrangle" such that a geometric figure surrounded by three straight lines is named a triangle. There are three remarkable features in the lesson plan. The first feature is the use of the different figures shown in Figure 4.4, including the intricate figures of Figure 4.4(c) and Figure 4.4(d), as teaching materials.

The second is the way the figures are shown to the students one by one in the order of Figure 4.4(a), Figure 4.4(b), Figure 4.4(c) and so on (see Figure 4.4) to encourage the students to remember the definition of a triangle and to use the definition in explaining explicitly the reason for their judgment. The third is the incorporated important question "Why would a figure with three connected points not be a triangle?" to shake the students' recognition of a triangle and to deepen their understanding of the definition of a triangle. It might be said that the three features are crystallized as a result of the collaborative team study on teaching materials before conducting the research lesson.

Reflective analysis of the research lesson

The research lesson was analysed using the detailed transcript (Maeda, 2014) and a photograph of the blackboard used in the lesson. As a result, the author identified the importance of emphasizing viewpoints of the figure and using counter-examples (Figure 4.4) and the different roles played by the teacher in two notable scenes of the lesson (Koyama, 2014, 2015). In the first notable scene, where Figure 4.4(c) was shown by the teacher, the students had opposing judgments about whether the figure is a triangle. After students talked for 1 minute in pairs, they actively exchanged their viewpoints of Figure 4.4(c) in a whole-classroom discussion. Some students presented their opinions in their own words. For example, one student said "Those who judge Figure 4.4(c) as not a triangle see the figure as a whole, including the spiral part. Meanwhile, those who judge Figure 4.4(c) as a triangle see the inside part of the figure surrounded by three straight lines".

During the classroom discussion, the teacher did not make any comments about the students' opinions and only wrote down keywords related to their

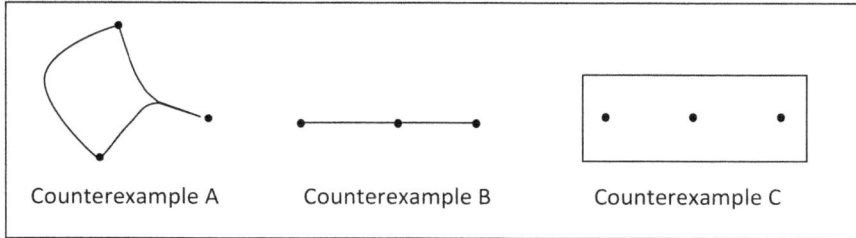

Figure 4.5 Counter-examples presented in the research lesson.

viewpoints of the figure on the blackboard; for example, "If we change our viewpoints", "It depends on the viewpoint of the figure", and "If we see the figure like this" In this scene, the teacher's supportive intervention functioned effectively for the students to exchange their judgements by referring to the definition of a triangle and to share different viewpoints of the same Figure 4.4(c).

The second notable scene in the research lesson relates to the discussion between the teacher and students. The scene began with the teacher's questioning of the students "Why would a figure with three connected points not be a triangle?" Immediately many students objected to what the teacher said. A student explained why a figure with three connected points might not be a triangle by drawing a figure (counter-example A) (see Figure 4.5) and saying, "If three points are connected by curved lines, the figure is not a triangle. Three points must be connected by straight lines".

Immediately the teacher counter-attacked again by saying to the students, "I see! Do you agree that a figure with three points connected by straight lines is a triangle?" Another student did not agree with the teacher and refuted what the teacher said by drawing a figure (counter-example B) with three points on the same straight line and saying, "I do not agree. If three points connected by a straight line is a triangle, then this (counter-example B) should be a triangle. Therefore, the triangle must be a figure encircling three points by straight lines". The student insisted on the importance of the figure being encircled by the straight lines. At that moment, many students seemed satisfied with their peer's refutation.

However, the teacher with a smile drew another figure (counter-example C) on the blackboard. In this scene, it can be said that the teacher's active intervention functioned effectively for the students to deepen their understanding of the definition of a triangle through a whole-classroom discussion with counter-examples.

Suggestions for improving performance in mathematics

Figure 4.6 summarizes the author's reflective analysis of the research lesson. As a result of the analysis, we recognize the importance of investigating

Three Proposals	Lesson plan	Data
P1: Viewpoint of figure P2: Counterexample P3: Role of teacher	Actual occurrence	D1: Detail protocol D2: Pictures D3: Field notes D4: Reflection by teacher

P1: Viewpoint of fig

Awareness of different viewpoints of the figure

P3: Role of teacher

Supportive role for students to exchange their judgments referring to the definition of triangle and to be aware of different viewpoints of figure

- In the first notable scene where Fig.4.4(c) was shown by the teacher, the students had opposite judgments about whether the figure is a triangle or not.
- After one-minute talking in pairs, they exchanged viewpoints on the figure in whole class discussion. "Those who judged Fig.4.4(c) not to be a triangle see the figure as a whole including part of spiral; those who saw Fig.4.4(c) as a triangle see the inside part of the figure surrounded by three straight lines".
- During the classroom discussion, the teacher did not make comments about the students' opinions and presentations but wrote on the blackboard only key words related to the figure, such as, "if we change our viewpoints …", "It depends on the viewpoint of the figure …" and "If we see the figure like this …".

(c)

P2: Counterexample

Teacher's key-questioning, "Why do you not say that the figure connected by three points is a triangle?" in order to question the students' recognition of a triangle and to deepen their understanding of the definition of a triangle.

P3: Role of teacher

Active intervention for students to deepen their understanding of the definition of a triangle through discussion with counterexamples.

- The second scene began with the teacher's questioning to the students, "Why do you not say that the figure connected with three points is a triangle?"
- Immediately many students objected to what the teacher said.
- A student explained the reason why the figure connected by three points is not a triangle by drawing counterexample A. "If the figure is surrounded by three points that does not make a triangle. Three points must be connected by straight lines".
- Then the teacher counterattacked, saying to the students, "I see. Do you agree that the figure connecting three points on the same straight line is a triangle?"
- Another student did not agree with the teacher and refuted by drawing the figure of counterexample B with three points on the same straight line. "I do not agree. If three points connected by the same straight line was a triangle, then this should be a triangle. Therefore a triangle must be the figure surrounding three points by straight lines".
- However, the teacher, with a smile, drew a figure (counterexample C) on the board.

Counterexample A

Counterexample B

Counterexample C

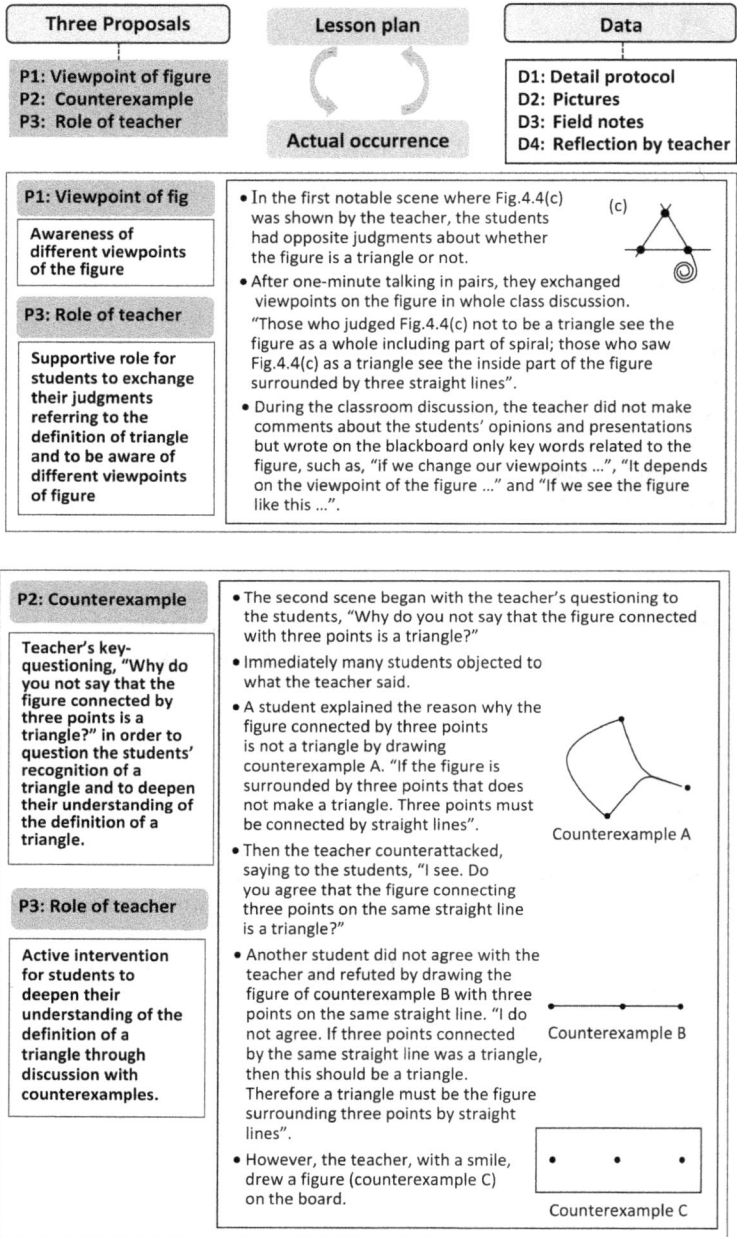

Figure 4.6 Summary of the author's reflective analysis of the research lesson.

mathematical teaching materials including teacher's questioning and roles in the classroom for developing students' mathematics thinking and their ability of mathematical problem solving.

As a result of the reflection on the lesson study at the HUAPS by focusing on the relationship between collaboration and reflection, the author insists that two types of complementary reflection function dialectically in the overall process of lesson study. The author calls them collaborative reflection and individual reflection. Collaborative reflection functions when a group of people reflect collaboratively on what the group did. Meanwhile, individual reflection functions when a person reflects individually on what the person did. The author (Koyama, 2019) proposes the framework of the dynamic cycle for the professional development of mathematics teachers and educators in the lesson study of school mathematics (see Figure 4.7).

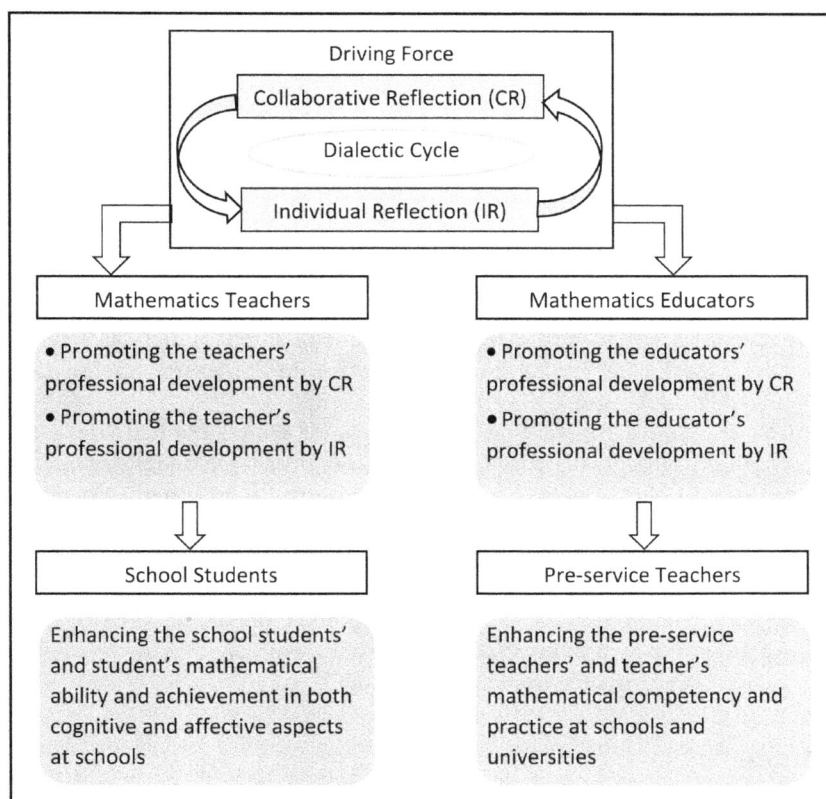

Figure 4.7 Framework of the dynamic cycle for the professional development of mathematics teachers and educators in the lesson study of school mathematics (Koyama, 2019, p. 45).

The dynamic cycle is driven by the dialectic cycle of two complementary reflections in lesson study for the professional development of teachers, which may improve both cognitive and affective aspects of the mathematical abilities and achievement of students. This aspect is represented on the left of Figure 4.7, with the two complementary reflections put as the driving force in lesson study. Meanwhile, the right side of Figure 4.7 implies that the dialectic cycle promotes the professional development of mathematics educators and greatly improves the mathematical competency and teaching practice of pre-service teachers at university and teachers at schools (Koyama, 2019, pp. 45–46).

Final remarks

In this chapter, the author shared the Japanese methods of problem-solving approach and lesson study for developing students' mathematics thinking and their ability of mathematical problem solving and for promoting professional development of mathematics teachers and educators.

As a result of reflective analysis of lesson study of problem-solving lesson, for developing students' mathematics thinking and their ability of mathematical problem solving, we are suggested the importance of investigating mathematical teaching materials used as learning materials for students, including teacher's questioning with counter-examples to shape students' recognition and teacher's flexible roles in the classroom.

And the author proposes a framework of a dynamic cycle driven by the dialectic cycle of two complementary reflections in the lesson study of school mathematics for the professional development of mathematics teachers and educators, which may improve both cognitive and affective aspects of the mathematical ability and achievement of students.

Finally the author insists that lesson study of school mathematics is not only an important means for the continuous professional development of mathematics teachers but also an authentic research area in the science of mathematics education to integrate research and practice related to school mathematics. Therefore, in further research, it is a critical issue for all of us in mathematics education to certify how the dynamic cycle driven by the dialectic cycle of two complementary reflections in the lesson study of school mathematics can function effectively and productively in improving students' and teachers' performance in mathematics.

Endnote

This article is the modified version of Koyama (2019) for the presentation at the Mico International Mathematics Teaching Summit held in Kingston, Jamaica, March 25-27, 2019. This work was supported by JSPS KAKENHI under grant number JP18H01016. Any opinions, findings and conclusions or recommendations are those of the author and do not necessarily reflect the views of the JSPS.

References

Becker, J. P., & Shimada, S. (Eds.) (1997). *The open-ended approach: A new proposal for teaching mathematics.* Virginia, USA, National Council of Teachers of Mathematics.

Burghes, D., & Robinson, D. (2009). *Lesson study: Enhancing mathematics teaching and learning.* London, UK, CfBT Education Trust.

Corey, D. L., Peterson, B. E., Lewis, B. M., & Bukarau, J. (2010). Are there any places that students use their heads?: Principles of high-quality Japanese mathematics instruction. *Journal of Research in Mathematics Education, 41,* 438–478.

Isoda, M., Stephens, M., Ohara, Y., & Miyakawa, T. (Eds.). (2007). *Japanese lesson study in mathematics.* Singapore, World Scientific.

Koyama, M. (2008a). Current issues impacting on mathematics education. *Bulletin of Graduate School of Education, Hiroshima University, Part II, 57,* 29–38.

Koyama, M. (2008b). Mathematics teacher training in Japan. In Burghes, D. (Ed.), *International comparative study in mathematics teacher training* (pp. 26–28). London, UK, CfBT Education Trust.

Koyama, M. (2010). Mathematics curriculum in Japan. In Leung, F. K. S. & Li, Y. (Eds.), *Reforms and issues in school mathematics in East Asia: Sharing and understanding mathematics education policies and practices* (pp. 59–78). The Netherlands, Sense Publishers.

Koyama, M. (2012). Examples of good practice in mathematics teaching and learning in Japan. In D. Burghes (Ed.), *Enhancing primary mathematics teaching and learning: Research report* (pp. 16–24). London, UK: CfBT Education Trust.

Koyama, M. (2014). The importance of viewpoints, counterexamples, and roles of mathematics teacher in the teaching and learning of geometric figures in primary school mathematics. *School Education, 1157,* 44–49. (In Japanese).

Koyama, M. (2015). Seeking high quality collaboration and reflection: A lesson study on primary school mathematics. In C. Vistro-Yu (Ed.), *In the pursuit of quality mathematics education for all: Proceedings of the 7th ICMI - East Asia regional conference on mathematics education* (Vol.1, pp.105–116). The Philippines: Philippine Council of Mathematics Teacher Educator (MATHED), Inc.

Koyama, M. (2019). Framework of a dynamic cycle for promoting the professional development of mathematics teachers and educators in the lesson study of school mathematics. *Hiroshima Journal of Mathematics Education, 12,* 33–48.

Lewis, C. (2002). *Lesson study: A handbook of teacher-led instructional improvement.* Philadelphia, USA: Research for Better Schools.

Maeda, K. (2014). Education of geometric figures for fostering students' logical thinking in primary school mathematics. *School Education, 1157,* 36–43. (In Japanese)

Ministry of Education, Culture, Sports, Science and Technology. (2008a). *Guidebook for the primary school mathematics in the Course of Study (2008).* Tokyo, Japan: Toyokanshuppansha Publisher. (In Japanese)

Ministry of Education, Culture, Sports, Science and Technology. (2008b). *Guidebook for the lower secondary school mathematics in the Course of Study (2008).* Tokyo, Japan: Kyoikushuppan Publisher. (In Japanese)

Ministry of Education, Culture, Sports, Science and Technology. (2009). *Guidebook for the upper secondary school mathematics in the Course of Study (2009).* Tokyo, Japan: Jikkyoshuppan Publisher. (In Japanese)

National Association for the Study of Educational Methods. (Ed.). (2011). *Lesson study in Japan.* Hiroshima, Japan: Keisuisha.

Polya, G. (1945). *How to solve it*. New Jersey, USA: Princeton University Press.

Sawada, T. (1997). Developing lesson plans. In J. P. Becker, & S. Shimada (Eds.), *The open-ended approach: A new proposal for teaching mathematics* (pp.23–35). Virginia, USA: National Council of Teachers of Mathematics.

Shimizu, Y. (2010). Mathematics teachers as learners: Professional development of mathematics teachers in Japan. In F. K. S. Leung, & Y. Li (Eds.), *Reforms and issues in school mathematics in East Asia: Sharing and understanding mathematics education policies and practices* (pp.169–179). The Netherlands: Sense Publishers.

Stigler, J., & Hiebert, J. (1999). *The teaching gap: Best ideas from the world's teachers for improving education in the classroom*. New York, USA: Free Press.

Takahashi, A. (2010). Lesson study: An introduction. In Y. Shimizu, Y. Sekiguchi, & K. Hino (Eds.), *Proceedings of the 5th East Asia regional conference on mathematics education* (Vol.1, pp.169–175). Tokyo, Japan: Japan Society of Mathematical Education.

Takahashi, A. (2014). Supporting the effective implementation of a new mathematics curriculum: A Case study of school-based lesson study at a Japanese public elementary school. In Y. Li, & G. Lappan (Eds.), *Mathematics curriculum in school education* (pp.417–441). The Netherlands: Springer.

Chapter 5

Uncovering everyday mathematics as a vehicle for equity: Investigating the funds of knowledge of diverse communities

Jodie Hunter and Rachel Restani

Background information

Both in the Pacific region of the world and internationally, the use of contextual mathematical tasks and problems has gained increased attention (e.g., Clarke & Roche, 2017; Reinke 2019; Wernet 2017). Potentially the use of contextually based tasks can engage and motivate students while also supporting them to make connections between their home and school use of mathematics. However, if schooling practices and contexts used in the mathematics classroom privilege particular culturally embedded ways of knowing, this then creates barriers to accessing education for students from diverse cultural groups (Joves, Siques, & Esteban-Guitart, 2015 Zipin, 2009).

Equity in our schooling systems across different nations and cultural contexts can only be achieved by explicitly connecting to and building on cultural, social, and linguistic contexts of students from non-dominant groups (Meaney, Trinick, & Fairhall, 2013; Parker, Bartell, & Novak, 2017; Planas & Civil, 2009). This includes understanding the identities of learners and the "funds of knowledge" they acquire in their lives the outside school (González, Andrade, Civil, & Moll, 2001; Moll, Amanti, Neff, & Gonzalez, 1992; Spiller, 2012).

Mathematics is a subject that as Presmeg (2007) describes has been for many years viewed as "culture-free". This results in a narrow view of mathematics as contained only within school settings and positions it as a subject where only specific groups of people from dominant cultures can succeed. This is reflected by the comments from an initial interview with students in the study reported in this chapter. The students when asked how they felt in relation to their cultural background in the mathematics classroom, positioned their cultural background within a deficit framing. For example, Mele shared her perception that "different cultured kids...they are smarter" and that "us, Niuean kids, we don't understand".

Similarly, Iosefa of Samoan heritage stated that "Samoa is different there with the maths. The Chinese people and the Japanese people, they are the smartest people in the world for maths". Interestingly, one of the students recognised the common deficit framing, saying "that stereotype is common that says Islanders don't really do maths at all". Across the world, to disrupt ongoing patterns of under-participation and achievement within schooling systems and Eurocentric views of mathematics and who can "do" mathematics requires that we provide ways in which to document and value the mathematical funds of knowledge of diverse communities in both Western and non-Western contexts.

The context of one small Pacific nation

Niue is a small self-governing island nation in the Pacific which is classified as a state in free association within the realm of New Zealand. It has a diverse population of around 1600 people. These include around 80% Niueans or part Niueans, 12% European and Asian and 8% from other close regional Pacific Island nations. The values and beliefs of these Pacific Island nations people most often reflect those within a collectivist society where family extends beyond the nuclear to include wider members of the local community.

Niue has two schools; a primary school for students aged 5–10 years and an intermediate/high school for students aged 11–17 years. The education system draws on the New Zealand curriculum and frequently New Zealand resources are used for teaching tools. Potentially, this means that the cultural knowledge and experiences of Niuean and other Pacific learners is excluded from the classroom. We suggest that given the very different lifestyle and experiences of students living in Niue, these students are being marginalised through reliance on contexts alien to their own cultural experiences. We argue that it is important that in contrast, the cultural knowledge and ways being of these learners are developed and built on within their classroom and schooling experiences.

At present, there have been limited studies specifically investigating funds of knowledge in mathematics in Pacific Island nations. Specifically, this chapter will examine the out of school experiences linked to mathematics, which this group of students living in Niue documented through photos. We argue that these photographs and subsequent interviews are a first step of documenting some of the funds of knowledge related to mathematics from a small Pacific Island nation.

Funds of knowledge

This chapter sets out from the contention that mathematics is neither culture free nor value free. By looking through a lens of everyday mathematics, at practices engaged in as part of everyday life in the home and community, we are able to identify the funds of knowledge of diverse groups of people (Civil, 2016). The funds of knowledge theoretical framing recognises that all people and cultures have bodies of knowledge and skills, which are historically accumulated, culturally developed and support individual/household functioning and well-being (Moll et al., 1992). Funds of knowledge is often associated with adult family practices and activity. A further development is "funds of identity" theory which recognises that dependent on learning experiences and trajectories, students may draw on family funds of knowledge and/or construct their own funds of knowledge (Esteban-Guitart, 2012; Moll, 2014). Both funds of knowledge and funds of identity attend to the dynamic complexities of peoples' lives by evidencing that there are many ways to be part of a cultural group.

International studies (e.g., Takeuchi, 2018; Williams et al., 2016) exploring funds of knowledge related to mathematics of non-dominant communities often involve interviews with parents to establish mathematics and literacy activities present in household routines. One key finding across studies is that parents frequently have initial difficulties identifying household activities that involve mathematical learning outside of cooking and construction. Importantly, studies confer that although parents are active in supporting their students to think about mathematical content, they undervalue their knowledge or role in students' mathematical learning.

Other studies (e.g., Moje et al., 2004; Zipin, Sellar, & Hattam, 2012) broaden the traditional approach of funds of knowledge by actively incorporating students into the process by asking them to document their funds of knowledge. These studies provide evidence for the usefulness of tools including cameras, social media apps and mobile devices to enable participants to document and share funds of knowledge in their everyday lives.

Researchers note that this enables students to both capture and share a wide range of experience including both unconventional and implicit scientific knowledge. Importantly, these studies suggest that supporting students to document their funds of knowledge related to content areas (e.g., science, mathematics) can potentially facilitate them to more readily recognise out-of-school STEM (Science, Technology, Engineering and Mathematics) experiences.

Previous studies (e.g., Si'ilata, Samu, & Siteine, 2018; Cooper & Hedges, 2014; Dickie, 2011; Hogg 2016) involving people from Pacific Island nations and funds of knowledge have generally been undertaken in New Zealand and have largely related to literacy practices, early childhood settings or generalised curriculum areas. These studies have shown the linguistic and cultural resources of Pacific learners. Dickie (2011) trained primary-aged Samoan students as junior ethnographers to document their out of school literacy use. The findings showed that literacy use was common in multiple sites including family, and church and neighbourhood settings, but often there was conflict in relation to values between the different sites.

For example, popular culture (e.g., hip-hop music) within neighbourhood sites conflicted with family and church values. Si'ilata, Samu, and Siteine (2018) used case studies to illuminate how Pacific student progress was accelerated when their teachers drew on their funds of knowledge. Making links between home and school contexts is critical if Pāsifika communities are to see mathematics as an important and lived aspect of their lives. As Havelková (2013) says, if education does not match what society considers important, students are likely to regard it as irrelevant to their lives.

To enable educators to develop appropriate contextually based mathematical tasks, we need to firstly identify both the activities and artefacts, which encompass everyday mathematics for diverse learners.

Methods

The data in this chapter is drawn from a wider study that examined the mathematical funds of knowledge of Pacific students both from Niue and New Zealand and how this could be used to develop contextual mathematical tasks. This chapter specifically focuses on 17 students and their families from six classrooms across two schools in Niue. It reports on the students' initial experiences of the use of contextual tasks in their classrooms and analyses their mathematical funds of knowledge. The students were aged between 9 years and 15 years old and there were nine girls and eight boys involved. All students were of Pacific descent with their cultural background shown in Table 5.1.

An initial workshop was held for students and their families to introduce the purpose and structure of the project. The first activity during the workshop was an examination and discussion of photographs depicting everyday activities that potentially involved mathematics. These included Samoan dancing, rugby, a Cook Island tivaevae (handmade quilt) and otai (a watermelon-based drink). Subsequently, the students and their families were asked to brainstorm and discuss other everyday activities that included mathematics. This activity was used to begin to alert participants to how mathematics is part of everyday life. Following this, cameras were provided to each family and they were asked to take photos of the mathematics they noticed in their everyday lives and any activities they did that involved mathematics.

Table 5.1 Participants' cultural background

Cultural background	Number of students
Niuean	7
Samoan	1
Niuean and Samoan	2
Tongan, Cook Island, and Niuean	2
Tongan	1
Tuvaluan	1
Niuean and European	2
Fijian	1

Data was collected through individual interviews with the students. An initial interview was held to determine where they saw maths in their home and community, investigate their perceptions of mathematics in their culture and explore their experiences of contextual tasks in the classroom. The second and final interview was held nine weeks later and involved photo-elicitation interviews. During these interviews, the students were asked to describe the situation or artefact in each photo. The interviewer also asked students why they took that picture and what mathematics they noticed in the situation or artefact. If students were unclear on the mathematics, then the interviewer inquired into the details of the situation. For example, *"how many pegs are used to hang each piece of clothing on a washing line? How many coconuts do the pigs eat in one sitting?"* Pressing students to describe more details created opportunities for the researcher to understand the mathematics involved in their daily activities. The students were also asked to reflect on where they saw mathematics in their life.

Data analysis drew on different forms of thematic analysis (Boyatzis, 1998). Firstly, to analyse the initial interviews, an emergent inductive strategy was applied to allow the themes to emerge from the data itself (Creswell, 2018). Data generated from the photo-elicitation interviews was coded in a number of ways using a deductive strategy. Firstly, the data was coded using the themes developed by Moje and colleagues (2004) to examine the different funds of knowledge including family, community, peer and popular culture. Secondly, the data was examined to investigate the different strands of mathematics that were apparent in the activities, artefacts and responses from the students.

In the next section, we begin by examining student perspectives of the use of contextual mathematical tasks in their classrooms. Following this, we investigate the mathematical funds of knowledge of this group of learners from Niue and the potential to draw on familiar contexts to develop rich, challenging mathematical tasks.

Use of contextual mathematical tasks in the classroom

Potentially, the consistent use of tasks without context can lead to both learners disengaging from the subject and failing to see the relevance of mathematics in their lives (Boaler, 1993; Henningsen & Stein, 1997). Using contexts that are tokenistic or difficult for learners to understand causes further difficulties with engagement. It also creates barriers for students who have to make sense of the context while also trying to solve the mathematical task. Unknown contexts result in students being unable to draw on their everyday home and cultural-based knowledge of mathematics (Lubienski, 2000; Meaney et al., 2013). In contrast, use of appropriate and relevant contexts in mathematical tasks provides an opportunity for learners to both draw on and connect their everyday knowledge to solution strategies while also affirming cultural identity (Betts & Rosenberg, 2016; Boaler, 1993; Civil, 2007; Matthews, Watego, Cooper, & Baturo, 2005).

In the initial interviews with the students, there was variation across the six classrooms in relation to the use of contextual tasks for mathematics. A small group of students (3 out of 17) reported that their teachers did not use contextual tasks. Most students (10 out of 17) reported that their teachers used everyday contexts in their mathematical tasks. However, the examples that the students provided, demonstrated that these were generic and unrelated to the specific local and cultural contexts of Niue. For example, Mele stated: "*So it's like if Sam went shopping and bought 13 apples and then he went and bought five more, so we have to solve how many apples Sam got at the shop*". This illustrated the disconnect between the tasks and context of Niue. Apples are imported into the Island nation and are expensive so it is unlikely that parents would buy that volume of apples.

Another group of students (4 out of 17) stated that their teachers developed some problems that related to everyday life on Niue: "*She refers to our plantations and sometimes the cost of food here*". However, the number of students who reflected this was limited and overall, these interviews indicated that the rich abundance of local and cultural experiences aligning with mathematics in Niue were not used as a resource for developing contextual mathematical tasks.

The following section will describe the mathematical funds of knowledge of these students and their families linked to family, community and peer sources that were reflected in their photographs and photo-elicitation interviews

Family funds of knowledge related to mathematics

Most commonly, during the photo-elicitation interviews, the students shared photos of activities (58 out of 85) that reflected their family funds of knowledge. Similar to the seminal work by Moll and colleagues (1992) and later work by Moje et al. (2004), these home-based funds reflected the work done within the home to support the household functioning. The focus on family is not surprising given that these students experience a communal way of life.

Within this category, many of the photographs (20 out of 58) reflected the agriculture work undertaken by families on Niue – activities around food that are required to maintain a household. Firstly, many students shared photos of themselves and their families collecting food both from the land and the sea. For example, students talked about harvesting their talo (taro, a starchy root crop) plantation, or collecting eggs from the chickens at the back of their houses. Embedded within their explanations were opportunities for mathematics. For example, when relating the number and proportion of chickens required to have sufficient eggs for families (especially because the students mentioned how chickens often went and never came back or were eaten by dogs).

The reef and sea were other common sources of food including shell-fish, crab and fish. A local delicacy is Unga (coconut crab). A student shared a photo of an Unga and described how her father would go Unga hunting at night. Another student described spear fishing on the reef using gesture to indicate angles: "*there is a big gun and then it has a hook and then you just hold it back and you press the trigger when you see a fish…sometimes the fish are down there [indicates far from the reef] and you have to walk a bit closer and then [indicates moving the spear gun to a different angle]. Like this is the fish and this is the human [using hands to show positions and angles] but the fish is not that big, it's a small one [uses hand to show a different angle]*". Within this example, we can see his use of rich aspects of measurement and geometry to explain reef fishing.

Agricultural activities also encompassed gardening and planting and looking after animals. Elements of measurement were highlighted as the students described how they worked with their families to grow food. Mele discussed the rectangular plots and area needed for banana plantations: "*you need to know how much basically it needs for it to grow. If they're all crowded, they won't grow easily or they won't get bananas on them. How much space you need to use, like centimetres or metres, you need them about 8–10 metres away from each other*". Clearly, Mele understood and drew on a range of linear measurement concepts. These could provide a basis for relevant contextual tasks.

Measurement of time was another important factor in the management of family plantations. Talo is a staple food crop on Niue. Paul described planting and harvesting talo as a continuous process: "*it takes probably like seven or nine months. You keep planting but sometimes you plant it for a particular occasion like show days or ceremonies, hair-cutting*".

Everyday mathematics was also described as embedded within the household routines of feeding the pigs and chickens. For example, a student discussed the ratio of buckets of food to the number of pigs and how long bags of chicken feed lasted on average.

Other forms of family funds of knowledge were also shared by students (15 out of 58). These related to household routines of shopping, preparing and cooking food. Many students spoke specifically about the Niuean dishes that they prepared with their families and their traditional ways of cooking.

For example, Dante spoke about preparing an umu (earth-oven) and making takihi, a traditional dish. Specifically, he noted: "*you need the correct amount for the layers of talo, pawpaw, and coconut cream and then get the right number so you don't either overcook it or undercook it*". Similarly, both proportions and ratios were evident in Mele's description of making watermelon smoothies: "*we had to know how much watermelon to put in the blender and how much ice and how much milk and ice cream. We needed about twelve pieces of watermelon and we needed two scoops of ice cream and five ice cubes and two cups of milk*". As these students talked, it was clear that they were drawing on the concepts of proportions and ratio.

Measurement concepts, including volume and capacity, were also apparent in the students' descriptions of food preparation. Commonly, passengers flying from New Zealand to Niue bring chilly-bins with food. Likewise, traditional dishes are prepared in Niue and bought by passengers to family in New Zealand. Taoro described how he worked with his family to prepare nane (coconut porridge) to bring to family in New Zealand: *"cooking my Niuean porridge that was a lot of measurements. You have to get the right consistency, if you didn't put enough pia (starch), the consistency would be too liquid but if you put too much in, it would be too thick. You have to measure the right amount of pia"*. He then continued to talk about the measurement of capacity and volume: *"then putting it in the containers and filling the chilly bin with containers and weighing the chilly bin"*.

These students are describing proportionality and measurement in ways which link their own community constructed mathematical knowledge and show their recognition of the relevance of mathematics. In these responses, we can see opportunities to cover both strands of proportional reasoning and measurement in ways that would support the students to draw on their existing knowledge and recognise the relevance of mathematics.

Also reflected in the students' descriptions (13 out of 58) of their family funds of knowledge related to mathematics was craftwork and leisure activities. Litea described the process of making fou fou (head-bands), kahoa hihi (shell necklace) and kahoa maile (leaf necklace): *"it's called a maile and they are like seven strands and then you wrap that up. My Aunty told me to put my hand like this [indicates over the top] and it's the angle, like a triangle shape. It has seven strands of string"*. She continued to describe how she made a fou fou: *"I cut them in angles, it's like a diamond so it's shaped like a diamond and you cut it across the centre"*.

The responses provided insight into how the students had already well-developed concepts of how patterning and geometry are integrated into traditional craftwork. Litea acknowledged the history of these activities stating: *our ancestors wore them for dancing*.

Students also referred to the family funds of knowledge in visiting different sites around the island including walking sea tracks and visiting chasms: *"the metres from the top going down to the sea track* or *how far it is and how long does it take to walk to the chasm"*. Here, they directly drew on measurement concepts to describe both the distances and times these visits took.

Illustrated throughout this section are the strong family funds of knowledge related to mathematics, which were evident in the students' responses.

Community funds of knowledge related to mathematics

Family funds of knowledge extended beyond those related to their immediate family members. Representations of community funds of knowledge were the second most frequent category (19 out of 85) captured in the students' photographs and discussed during the interviews. These were community activities organised by people from the same village or family but with people from different villages across the island attending the event.

There were two different events commonly documented by the students (15 out of 19) as sources of everyday mathematics. Firstly, every month one Niuean village is calendared to hold an annual Show Day. The Show Day provides the village with opportunities to raise money and showcase their craftwork, local produce, and runs food stalls, games, and competitions. Some students discussed the preparation of food sold at the Show Day. This included both making the food, dividing it into plates to be sold, arranging the plates on a table and selling the food. In other responses they described working out parking arrangements and seating for tables. These illustrated their use of a wealth of mathematical concepts in arranging the event.

A group of students shared photographs of hair cutting and ear-piercing ceremonies. These are traditional coming of age ceremonies where for boys, the hair that has been grown since babyhood is ceremonially cut and for girls, their ears are ceremonially pierced. Guests are invited to the ceremony and the following feast. Monetary contributions are made to the child from the guests and the family make a reciprocal gesture of providing a basket of food for each family group attending the ceremony.

Again, the students described preparation of food for the feast. This included weaving and preparing baskets for the guests to take food home. For example, Litea described how food was given out after an ear-piercing ceremony: *"in the basket, they put talo, chicken, two packets of corned beef and half a pig with a head"*. Other students described the collection and counting of money that was gifted during the ceremony. These community events involved the students describing how they worked with numbers and operations, and a range of measurement concepts including area and volume.

Peer funds of knowledge related to mathematics

Only a small group (8 out of 85) of photographs and interview responses related to peer funds of knowledge in relation to mathematics. These predominantly focused on sports including running, weightlifting and rugby. For example, Mele made reference to measurement of mass and the use of number in relation to weightlifting: *"I had to put on thirty-five, so we had to put on two tens because the bar was fifteen and so we had to put on the two tens on the side to make it thirty five"*. Other students talked about measurement of distance and time in relation to running and having races with friends and siblings.

Discussion and conclusion

Clearly, the findings illustrate that the responses gathered from the students and their families through the use of photography and photo-elicitation interviews provide a valuable illustration of their rich use of mathematics in their daily lives. A large number of the students' contributions highlight the rich range of contexts in the students' lives that are authentically linked to mathematics.

Also evident is the specificity of the cultural knowledge and experiences and how these link to mathematics of the students living in Niue. Although this is an example of one small nation, the results are applicable to other diverse communities and nations around the world and suggest the need for similar studies to be done in other small (and not so small) diverse communities wherever they are located.

What is evident when reviewing the students' responses are the difficulties and challenges created by using mathematical resources from another cultural group (in this case New Zealand). For this collectivist society, it was clear that much of their mathematical activity was placed within family and community cultural and social activity, which contrasted with the school mathematics they experienced which disregarded this activity. This was shown in the initial interviews completed at the start of the study. In these interviews, many of the participants positioned their cultural background as deficient in relation to mathematics.

In contrast, in the final interviews, all students were able to speak confidently about the many ways that mathematics was part of their life. They identified its presence within all forms of family, community and peer funds of knowledge. As one student reflected on embarking on photographic exercise: *"at first, I thought it was going to be a bit hard…but once you have a mind-set of looking for maths, it's actually more common than you think"*.

Our findings match those of others who have undertaken similar work. These other studies also with diverse students (e.g., Moje et al., 2004; Zipin, Sellar, & Hattam, 2012) have provided evidence that positioning students to document their funds of knowledge can lead to them more readily recognising mathematics or science within their lives.

We argue that there is a critically important need to position diverse learners to explore their use of mathematics in their daily lives outside of schooling as a way to address their own deficit perceptions of themselves as mathematicians. However, such activity also provides a tool for educators to grow their understanding of the rich everyday use of mathematics of diverse communities.

Arguably, to create mathematics for sustainable economic growth and job creation in small nations like Niue, educators need to address issues of equity and access to mathematics. It is imperative that in developing curriculum, resources and mathematical tasks, we must consider the existing funds of knowledge of all learners. Importantly, by developing educators' understanding of the rich funds of knowledge of diverse communities, we can move towards developing more equitable classroom contexts and educational systems.

This chapter has used a case study of a group of students in Niue to highlight the mathematical experiences at home and in the community. These contrast sharply with the experiences of those from the dominant cultural groups within more individualistic societies who often develop curriculum and resources.

We contend that through aligning the contexts of mathematical tasks with the experiences of the students, learners are able to draw on their everyday home-based knowledge of mathematics and use it to grapple with high levels of challenging mathematics and thus attain more equitable achievement. At the same time, learners will be able to see the relevance of mathematics in their daily lives and want to continue studying it through the links they have made between the rich home use of mathematics and what they do at school.

References

Betts, P., & Rosenberg, S. (2016). Making sense of problem solving and productive struggle. *Delta-K, 53*(2), 26–31.

Boaler, J. (1993). The role of contexts in mathematics classrooms: Do they make mathematics more real? *For the Learning of Mathematics, 13*(2), 12–17.

Boyatzis, R. E. (1998). *Transforming qualitative information.* Thousand Oaks, CA, Sage.

Civil, M. (2016). STEM learning research through a funds of knowledge lens. *Cultural Studies of Science Education, 11*(1), 41–59.

Civil, M. (2007). Building on community knowledge: An avenue to equity in mathematics education. In N. Nassir & P. Cobb (Eds.). *Improving access to mathematics: Diversity and equity in the classroom* (pp. 105–117). New York, Teachers College Press.

Clarke, D., & Roche, A. (2017). Using contextualized tasks to engage students in meaningful and worthwhile mathematics learning. *The Journal of Mathematical Behaviour, 51*, 95–108. https://doi.org/10.1016/j.jmathb.2017.11.006

Cooper, M., & Hedges, H. (2014). Beyond participation: What we learned from hunter about collaboration with Pasifika students and families. *Contemporary Issues in Early Childhood, 15*(2), 165–175.

Creswell, J. W. (2018). *Research design: Qualitative and quantitative approaches* (5th ed.). Thousand Oaks, CA, Sage.

Dickie, J. (2011). Samoan students documenting their out-of-school literacies: An insider view of conflicting values. *Australian Journal of Language and Literacy, 34* (3), 247–259.

Esteban-Guitart, M. (2012). Towards a multi-methodological approach to identification of funds of identity, small stories and master narratives. *Narrative Inquiry, 22*, 173–180.

González, N., Andrade, R., Civil, M., & Moll, L. (2001). Bridging funds of distributed knowledge: Creating zones of practices in mathematics. *Journal of Education for Students Placed at Risk, 6*(1-2), 115–132.

Havelková, V. (2013). Jourdain effect and dynamic mathematics. In *Proceedings of the 10th international conference on efficiency and responsibility in education (ERIE 2013)* (pp. 182–188). Prague.

Henningsen, M., & Stein, M. K. (1997). Mathematical tasks and student cognition: Classroom based factors that support and inhibit high level mathematical thinking and reasoning. *Journal for Research in Mathematics Education, 28*(5), 524–549.

Hogg, L. (2016). Applying funds of knowledge theory in a New Zealand high school: New directions for pedagogical practice. *Teachers and Curriculum, 16*(1), 49–55.

Joves, P., Siques, C., & Esteban-Guitart, M. (2015). The incorporation of funds of knowledge and funds of identity of students and their families into educational practice. A case study from Catalonia, Spain. *Teaching and Teacher Education, 49*, 68–77.

Lubienski, S. T. (2000). Problem solving as a means toward mathematics for all: An exploratory look through a class lens. *Journal for Research in Mathematics Education, 31*(4), 454–482.

Matthews, C., Watego, L., Cooper, T. J., & Baturo, A. R. (2005). Does mathematics education in Australia devalue indigenous culture? Indigenous perspectives and non-indigenous reflections. In P. Clarkson, A. Downtown, D. Gronn, M. Horne, A. McDonough, R. Pierce, & A. Roche (Eds.). *Proceedings of the 28th annual conference of the mathematics education research group of Australasia* (pp. 513–520). Melbourne, Vic, University of Melbourne.

Meaney, T., Trinick, T., & Fairhall, U. (2013). One size does not fit all: Achieving equity in māori mathematics classrooms. *Journal of Research in Mathematics Education, 44*(1), 235–263.

Moje, E. B., Ciechanowski, K. M., Kramer, K., Ellis, L., Carrillo, R., & Collazo, T. (2004). Working toward third space in content area literacy: An examination of everyday funds of knowledge and discourse. *Reading Research Quarterly, 39*(1), 38–70.

Moll, L. (2014). *L. S. Vygotsky and education.* New York, Routledge.

Moll, L. C., Amanti, C., Neff, D., & Gonzalez, N. (1992). Funds of knowledge for teaching: Using a qualitative approach to connect homes and classrooms. *Theory into Practice, 31*, 132–141.

Parker, F., Bartell, T., & Novak, J. (2017). Developing culturally responsive mathematics teachers: Secondary teachers' evolving conceptions of knowing students. *Journal of Mathematics Teacher Education, 20*, 385–407.

Planas, N., & Civil, M. (2009). Working with mathematics and immigrant students: An empowerment perspective. *Journal of Mathematics Teacher Education, 12*(6), 391–409.

Presmeg, N. (2007). The role of culture in teaching and learning mathematics. In F. K. Lester (Ed.). *Second handbook of research on mathematics teaching and learning* (pp. 435–458). Greenwich, CT, Information Age Publishing.

Reinke, L. T. (2019). Toward an analytical framework for contextual problem-based mathematics instruction. *Mathematical Thinking and Learning, 21*(4), 265–284.

Si'ilata, R., Samu, T., & Siteine, A. (2018). The va'atele framework: Redefining and transforming Pasifika education. In E. A. McKinley & L. T. Smith (Eds.). *Handbook of Indigenous Education* (pp. 1–26) Singapore: Springer.

Spiller, L. (2012). How can we teach them when they won't listen? How teacher beliefs about Pāsifika values and Pāsifika ways of learning affect student behaviour. *Set: Research Information for Teachers, 3*, 58–67.

Takeuchi, M. A. (2018). Conversions for life: Transnational families' mathematical funds of knowledge. In T. G. Bartell (Ed.). *Toward equity and social justice in mathematics education* (pp. 127–143). Switzerland, Springer.

Wernet, J. W. (2017). Classroom interactions around problem contexts and task authenticity in middle school mathematics. *Mathematical Thinking and Learning: An International Journal, 19*(2), 69–94.

Williams, J. J., Tunks, J., Gonzalez-Carriedo, R., Faulkenberry, E., & Middlemiss, W. (2016). Supporting mathematics understanding through funds of knowledge. *Urban Education, 55*(3), 476–502. DOI: 10.1177/0042085916654523.

Zipin, L. (2009). Dark funds of knowledge, deep funds of pedagogy: Exploring boundaries between lifeworlds and schools. *Discourse: Studies in the Cultural Politics of Education, 30*(3), 317–331.

Zipin, L., Sellar, S. & Hattam, R. (2012). Countering and exceeding 'capital': A 'funds of knowledge' approach to re-imagining community. *Discourse Studies in the Cultural Politics of Education, 33*(2):179–192. https://doi.org/10.1080/01596306.2012.666074.

Zipin, L., Sellar, S. & Hattam, R. (2012). Countering and exceeding 'capital': A 'funds of knowledge' approach to re-imagining community. *Discourse Studies in the Cultural Politics of Education, 33*(2): 179–192. https://doi.org/10.1080/01596306.2012.666074.

Chapter 6

Mathematical progress with Pāsifika learners

Roberta Hunter

Context

Although many Pāsifika students enter schools in Aotearoa/New Zealand with deep and rich backgrounds of understandings and experiences within their own cultural and social groups, a very large number of them quickly become underachievers in the schooling system. This has been an enduring pattern over many decades, more generally in education, and specifically in mathematics.

The reasons are many and varied, but central to their consistent pattern of failure in the Aotearoa/New Zealand education system are the structural inequities they constantly meet. These barriers cause them cultural and social dissonance and they quickly develop a view of mathematics as having little relevance or connection to their lives, or as something unattainable (Hunter & Hunter, 2017).

For a considerable length of time, we have been engaged in research to reverse the low trends in Pāsifika achievement in mathematics. In this chapter, we aim to illustrate how one aspect of a transformative school-based professional learning and development research approach Developing Mathematical Inquiry Communities (DMIC) (for more detail see Hunter, Hunter, Anthony, & McChesney, 2018) provided more equitable outcomes for these Pāsifika students. DMIC is set within a culturally sustaining model (Paris, 2012) and draws on what Kazemi and her colleagues (2009) describe as ambitious mathematics.

What we want to illustrate is how the cultural values of a collectivist group of people, when understood and honoured in mathematics classrooms, hold huge potential to support more equitable outcomes and achievement for Pāsifika students. However, our focus is not just on school achievement.

Therefore, the intent of this chapter is to show how teachers building on the cultural values of this diverse group of learners, can construct classroom social and socio-mathematical norms which change the mathematical opportunities for the students. Through doing so Pāsifika students can be supported to develop a strong mathematical disposition while at the same time maintain a positive cultural identity as a Pāsifika learner.

To start, in the next section we describe who this group of students are, and explain why they might struggle in Aotearoa/New Zealand mathematics classrooms—classrooms which best reflect the cultural capital (Bourdieu & Passeron, 1973) of the mainstream groups predominantly in our classrooms.

Pāsifika Nations people in New Zealand

Pāsifika Nations people have held an important place in the history and shaping of Aotearoa/New Zealand for the past century. This goes back beyond both the political and economic contributions they make to society to include their ancestral links with Māori, the indigenous people of this country. These links help shape the rich social and cultural tapestry of Aotearoa/New Zealand as it is today.

As first cousins to Māori but as migrant groups to Aotearoa/New Zealand, they are composed of multiple generations within the country, as well comprising diverse and indigenous groups of Island Nations people, each with their own language and cultural and social ways of being (Hunter & Hunter, 2017). Nevertheless, across these Island Nations their communal approaches to how their society functions shapes all their cultural values and provides a common thread between Māori and the different Pāsifika Nation's people.

The cultural values of Pāsifika Nations people are constructed within, and shaped by, their collectivist and communal ways of life. These make a sharp contrast with the more individualistic approach, and the values which sit within this approach prevalent within the dominant and mainstream Pākehā (European) group (Hunter & Hunter, 2017) and which shape the educational practices in Aotearoa/New Zealand. These practices that are assumed to be normal for everyone, closely match with those of the dominant group of Pākehā learners in mathematics classrooms. However, their use positions Pāsifika learners unfavourable.

This negative positioning results in their disengagement from mathematics, but also maintains their on-going low levels of achievement. Most importantly, it causes them to develop an inferiority view of themselves and their own ethnic and cultural groups, as mathematicians (Hunter & Hunter, 2017). If, as Gutiérrez (2002) argues we should consider participation as important as achievement then these Pāsifika students are experiencing a gross disservice. In this chapter, we focus on only one aspect of the professional learning and development research within DMIC and show what happens when teachers explicitly draw on and structure classroom interaction patterns in ways that allow Pāsifika students participation in accessing high level and challenging mathematical activity embedded within a range of mathematical practices (RAND Mathematics Study Panel, 2003).

Mathematical practices

It is important to engage in a set of mathematical practices in construction of a mathematical disposition (RAND Mathematics Study Panel, 2003). These practices emerge through the interactive discourse students participate in and extend beyond mathematical knowledge to include particular actions and ways of using and doing mathematics (Ball & Bass, 2003). Many examples of mathematical practices abound and we have written about them extensively in other papers (see Civil & Hunter, 2015; Hunter, 2008, 2013;; Hunter et al., 2018). We describe them as practices which efficient users of mathematics use when solving problems. These include such examples as explaining, justifying, representing or generalising mathematical reasoning. What is important to the construction and use of mathematical practices are particular modes of communicating reasoning (Skilling et al., 2016). This includes asking questions to ensure deeper understandings as well as engaging in challenging the reasoning of others.

We want to show in this chapter how teachers can use scaffolding methods for Pāsifika and other diverse students to engage in robust discursive discourse premised within a set of proficient mathematical practices. At the same time, we want to explain how these students can maintain their cultural integrity while building these skills through use of Pāsifika values set within a collectivist society.

Pāsifika values

Within a collectivist society, notions of fono (family and as an extended rather than nuclear group) are central to how the values are considered. Within our own research, we have extensive data which illustrates how important fono are to everything the students do in their lives including within the school setting (see Civil & Hunter, 2015; Hill, Hunter, & Hunter, 2019; Hunter et al., 2016).

Woven within the values of what it means to be a member of fono are such concepts as reciprocity, respect, service, inclusion, relationships, spirituality, leadership, collectivism, love and belonging (Anae et al., 2001). Some people might argue that these values are appropriately applied to most people in all society. However, there are many subtle, and not so subtle differences between how they are enacted within either a collectivist or individualistic society. These differences in meaning are at the basis of considerable dissonance for many Pāsifika students as they try to navigate a school world, which conflicts with many of their core beliefs founded within the values they have been raised with in their home world.

In many studies (e.g., Hunter, 2008, 2013; Hunter & Hunter, 2017), we have described how Pāsifika students are more reticent to talk than those from the more dominant cultural groups within Aotearoa/New Zealand. In Hunter and Hunter (2017), we suggested that this was probably a major cause for the disproportionate placement of Pāsifika students in low ability set groups. Set grouping has been a commonplace practice in Aotearoa/New Zealand for many years with many negative outcomes for Pāsifika learners.

For example, we know that the placement of Pāsifika students in low ability groups supports teachers developing deficit views and low expectations (Rubie-Davies, 2016). We also know that the practice of placing Pāsifika students in ability groups conflicts with the beliefs and values of many Pāsifika Nations people. This is because these grouping structures encourage competitiveness and place importance on the individual rather than group success. This emphasis is in complete contrast to notions of communalism and collectivism where the individual's success is judged by the success of the collective as a whole. Here Pāsifika notions of respect and service are important to consider because the role of the individual within the context of a mathematics group is to ensure that mathematical understandings are constructed and understood collectively.

In other papers (e.g., Civil & Hunter, 2015; Hunter & Anthony, 2011; Hunter, Hunter, & Bills, 2019) we described some of the dissonance caused by the clash in values many Pāsifika students experience within mathematics classrooms in Aotearoa/New Zealand. In Hunter and Anthony (2011), we outlined how many of the Pāsifika students considered that they learnt mathematics best through passive listening to teachers. Passive listening rather than active participation and inquiry align with the Pāsifika value of respect.

Teachers are thought to be elders and considered to have all the knowledge that is always considered to be right and unchallengeable. In the same article, we wrote about how these same students also described their discomfort at needing to question and challenge peers about their mathematical reasoning because they saw it as being disrespectful. Nevertheless, learning mathematics includes learning how to talk and act as mathematicians (Boaler 2016). In this chapter, we want to show how teachers positioned the students to talk in ways that respected and honoured their cultural values while at the same time they learnt to work as mathematicians.

In the next section, we provide a short summary of the methodology we used.

Methodology

The research reported on in this chapter is part of a larger ongoing longitudinal project that spans more than 10 years. The participants include approximately 300 teachers in 27 Aotearoa/New Zealand urban primary schools. The students predominantly come from very low socio-economic home environments and are predominantly of Pāsifika Nation groupings.

These students may be immigrants or range across first-, second- and third-generation born to Aotearoa/New Zealand. The teachers come from diverse backgrounds although few are of Pāsifika ethnicity. Their length of teaching experience varies although predominantly they tend to be less experienced or overseas trained as is typical in many of our high poverty schools. The teachers reported on in this paper had been in the project more than a year and were selected as representative of the common practices observed in the classrooms.

Data for this chapter drew on video-recorded classroom observations, field notes and teacher interviews. Lessons were often followed by teacher interviews. The interviews included the use of a set of open-ended questions, which allowed for multiple responses.

Analysis of the data consisted of comparing and contrasting responses from the different teachers, developing codes and identifying and categorising patterns and themes. When emerging themes and patterns were identified as consistent across the different participants they were explored further. For this paper, consistent themes were used to represent the actions the teachers took to develop classrooms in which all the students actively participated in mathematical activity.

Findings and discussion

In all classrooms we report on in this chapter the teachers had, over the past year or more, shifted their practices from the more traditional forms of direct instruction towards using the tenets of ambitious mathematics (Kazemi, Franke, & Lampert, 2009) and culturally sustaining practices (Paris, 2012). This entailed the use of more open and flexible pedagogy built around the teacher noticing and responding to student reasoning and participation. The students were given high level, challenging, group-worthy tasks (Featherstone et al., 2011). These were written within known Pāsifika contexts so that the students did not have to struggle with both the context and the mathematics. They were also written to draw on a range of skills to ensure as many opportunities as possible for group members to participate.

The classroom lessons followed what Smith and Stein (2011) have suggested as best practice to promote productive discourse. Each lesson followed a similar format in that a problem was launched by the teacher and the students then worked in groups of four to construct shared problem solutions. As they worked in their small groups, the teacher monitored both the emerging mathematical reasoning and the patterns of participation. Each lesson ended with a large group sharing session where the teacher sequenced student explanations and facilitated discussion, inquiry and argumentation. The lesson concluded with the teacher connecting the different mathematical thinking to a big mathematical idea.

Developing a safe supportive learning environment

A theme that emerged immediately across all the teacher interviews was their need to develop a safe and supportive learning environment. This was seen as a clear issue of equity and was seen as a key task for these teachers of Pāsifika students. They described the necessity for these students to feel safe and secure if they were to engage in social and intellectual risk-taking during mathematical activity. Central to all the actions teachers took was the framing of their pedagogical actions around the notion of the family as a collective, within the Pāsifika cultural values they were drawing on. As one teacher explained:

TEACHER: "I know for my Pāsifika students, family is everything. So the norms that I establish is that we are a family. The concept of family means to say that we are in this together. If someone doesn't understand, we are there to help her or him. Just like the context at home, if you struggle with something at home you ask a family member to help and build on that. That is how we work in the classroom. There is no one leader or individual in the group, it's about us…it's about the reciprocal learning, the sharing of ideas."

The way in which families within the Pāsifika Nations people extend beyond the immediate family to include all other members of the village (and in New Zealand, e.g., neighbours and other people in the local community or other church goers) was also drawn on. This was extended past just the need to collaborate and used to embody the ways in which construction of a mathematical explanation and justification depended on all members' contributions. This meant that ultimately no one individual in a group owned the explanation – rather the group collectively owned the responsibility for it and for each other's understandings.

For example, a teacher used a ratio problem within the context of making a ta'ovala (a Tongan ceremonial costume). The making of ta'ovala require considerable time and expertise on the part of different groups of people and he uses this as a metaphor for how he wants the students to work together on construction of their mathematical explanation. He begins by discussing how they are made:

TEACHER: "They are made in a group and it takes quite a long time. But if you work in a group how would that be of benefit?"
STUDENT 1: "It would be faster."

Not wanting to stop at the notion of many hands might lighten the load the teacher pressed further to ensure that the students realised the value of having multiple understandings shared.

TEACHER: "The weaving would get done faster. In our mathematics today we are going to be looking at a problem where people are weaving them together and thinking about how long it would take to do that. Think of the reasons why are we working in a group to solve our maths problems?"
STUDENT 2: "To share our ideas."
TEACHER: "To share our ideas. That's a good answer. Why else are we working in a maths group?"

The teacher honours the student's contribution by revoicing it then continues to press.

STUDENT 3: "Different ways of doing it."

TEACHER: "We have got different ways of doing it. We are sharing it, sharing it with other people in our group. How does that help us in our maths group?"

STUDENT 4: "So everyone understands."

TEACHER: "So we can share our ideas and our different ways of doing it so we can understand how different people did it. I might say oh I never thought that way but Hemi told me a good way to do it, which if you think back to the people making the ta'ovala I am sure when you first start doing it, it is hard but the more experienced people help you learn from them as you go and in the end it is all shared knowledge."

In this short discussion, the teacher has reinforced the need to respect the different contributions and capabilities within the mathematics groups. At the same time, he has affirmed that success is achieved through the collective working together within the Pāsifika concepts of inclusion and reciprocity.

Through this process, he has indicated, everyone is able to learn together and construct rich deep mathematical understandings. Clear evidence is provided of the teacher's awareness of culturally sustaining practices (Paris, 2012) while at the same time developing classroom social norms.

In other classrooms, the different teachers re-visited notions of the family and the strengths inherent in being a member of a family group. They made direct links to their students' family contexts and emphasised that family members take different roles all of which collectively make different cultural activities successful.

Anae and colleagues (2001) have shown the importance of family as central to the values and beliefs, which sustain Pāsifika communities. They drew on the different cultural activities their families might do together which were originally imported from their island nation but now were part of the fabric of New Zealand/Aotearoa Pāsifika communities. These ranged across the different island groups and supported their sense of pride in their own ethnic origins.

This included encouraging student voice from the quiet reticent members of the classroom community by building on the Pāsifika Nations students' own sense of cultural pride by reinforcing their need to represent themselves well as family members. This is illustrated in the statement of a teacher to a quietly spoken, shy student:

TEACHER: "Remember you are a member of our whanau (family) so you need to be loud and proud and confident…we are all ready to listen and think."

In taking this action, the teacher had removed the negative tone of the directive but at the same time had used notions of accountability to the collective in a positive way to press the student to explain. Classrooms which drew on the Pāsifika concept of family as an extended and wider group in mathematics lesson gave the students opportunities to transfer their home understandings of accountability to the collective into the school setting.

Developing individual and group accountability

Within ambitious mathematics classrooms, it is important that students are responsible for their own sense-making and the sense-making of others. Consistently, across the DMIC classrooms we observed that teacher actions placed emphasis on both individual and group accountability through emphasising the Pāsifika values of respect and service. In these classrooms, clear expectations were placed on the students that they had an individual responsibility to understand and make sense of the mathematical reasoning, and at the same time a collective responsibility that they made sure their peers understood this also. An example of this is outlined in the following excerpt. A teacher stops the students as they work in groups and says:

TEACHER: "I have just seen some really great work going on here. Tere was explaining his thinking and Mere asked him a question. That's great because Mere was making sure she knew what was going on. But then Tere did not leave it there because when he answered Mere's question he did not leave it there, he checked in that everybody else got it. That's like what we talked about when you are in a kapa haka group (a Māori cultural performance group). You can't leave anybody behind and you just keep working together until everyone is with you".

The concept of service within the Pāsifika values contrasts with its commonly accepted meaning in the Western world (as being something someone is expected to do for you and often rewarded in monetary ways). In this instance, the teacher is not only promoting the need for accountability but also reinforcing the concept of service as putting others before self and being accountable for them constructing both their personal and shared knowledge.

In another lesson, the same teacher drew the students' attention to the parallels between a group of students working together and the actions of a Pāsifika dance group. Rather than having any individual star of a dance group the ethos of these groups is to induct, support and challenge members until they achieve similar levels of expertise.

In this instance, she drew attention to similarities in the role of the tuakana (elder brother or other people of similar age) teina (elder sister or other people of similar age) in a cultural situation. She linked her observations to the individuals she saw taking leadership and actively supporting other group members to question and challenge the mathematical reasoning:

TEACHER: "There's really interesting korero (talk) going on…this group were having problems and arguments… and Wiremu you weren't giving out the answers and that's really good but you were pushing them to think. Yes, you had everyone talking about and discussing how they were going to sort out the ideas. You were challenging and other people were following your lead so the arguing was kapai (great)."

Pāsifika students find questioning and challenging others very difficult because it directly conflicts with their home and culturally developed concept of respect. This teacher has shown that she recognised the need to provide positive models of how students are to engage in inquiry and arguing to allow the student opportunities to develop their own models of how to achieve this respectfully. Her descriptions provide a sense of normalisation towards discursive interactions that are not just premised on agreeing but also involve disagreement as part of the construction of a mathematical explanation.

At the same time, she illustrated the way in which the tuakana had inducted others into the role of questioning and challenging. These actions parallel many described in previous papers (e.g., Hunter & Anthony, 2011; Hunter & Hunter, 2017) where we have shown the critical need to support students to learn safe ways to be accountable to themselves and others while engaging in discursive interactions. In addition, we want them to remain culturally safe and confident so that they can participate to achieve in ways Gutiérrez (2002) previously promoted.

Developing ways of questioning and challenging

A central component of constructing classrooms where all students capably engage in a range of mathematical practices is the mathematical discourse developed. It is essential that the students are able to readily question to extend mathematical explanations and also challenge so that justifications for mathematical reasoning are provided. Through engaging in mathematical argumentation many other practices are used including representing and generalising reasoning.

Within all the classrooms we observed, we noted the careful attention given to the development of ways all the Pāsifika students could engage in questioning and challenging others and so develop mathematical argumentation. In many instances, we observed how the teachers highlighted examples of when they saw students do this is a positive and productive manner.

TEACHER: "Wow Tahi, you really made them think but you did that with a clear question that challenged them to think again and check their explanation."

The teachers also provided explicit models of ways to engage in respectful questioning and challenging. This was important to provide the students with equitable opportunities to participate fully in what Ball and Bass (2003) described as discursive mathematical discourse that we know leads to deep and rich understandings.

Many teachers voiced that their goal was to enable them to construct an appropriate voice to engage in an ambitious mathematics classroom. Models of questions and challenges were displayed and role-played, and the teachers regularly took opportunities to outline the types of questions or statements they could use. For example, before the students started their small group activity the teacher said:

TEACHER: "You have got to ask those questions because that is what good mathematicians do. You have got to ask what did that mean or what did you do there in that bit. Keep asking those questions until you really understand. That is even more important when you disagree but you need to say why, use your maths but do it politely and ask those questions, or say I don't agree because, friendly argue so they have to justify their thinking."

In this excerpt, the teacher has drawn the students' attention to their need to use the voice of a mathematician while also pressing them to be accountable to constantly collaborate and sense-make. He has also laid out their right to disagree but this time he has made them accountable not to their peers but to the mathematics under consideration. At the same time, he has positioned what mathematicians do as agreeing and disagreeing within the term "politely."

Politeness is a key concept that underpins the Pāsifika concept within relationships, love and respect. The way in which the Pāsifika values were used enabled these students to engage in the productive discourse Ball and Bass (2003) describe.

Other evidence of how teachers built on the Pāsifika concept of respect is shown in actions where the teachers supported the students to build their confidence to engage in a range of mathematical practices. Speed is a common component of traditional classrooms but in these classrooms, the teachers slowed the pace down to allow extensive talking and thinking time.

Many Pāsifika meetings take the form of what we describe as talanoa. Rather than direct talk and questioning talanoa involves a circular discussion until a point is reached. No time component is involved.

The concept of time has a different meaning within Pāsifika cultural gatherings and elders are often seen to talk, pace in silence for a time undisturbed, and then talk again. It is considered disrespectful to question or interrupt these elders. The teachers drew on this behaviour and allowed extended wait time, prolonged silence and many opportunities for the students to rethink and reform their ideas.

For example, a teacher intercedes when a child hesitates and another child goes to talk and says:

TEACHER: "No, no, don't speak let her think and she can take as much time as she wants."

Humour was also readily accepted as a way to lighten the tension. Such actions were used as a way to open the space for less confident talkers. They also indicated to the students that their teachers realised that the new patterns of interaction required in the discursive interactions might not sit comfortably with them but that they had time to grow into them. Civil and Hunter (2015) also describe the importance of space, time and humour to allow diverse students to participate in ways they are comfortable with.

Conclusion and implications

Participating in and learning mathematics is about learning the key to how mathematics as a discipline works. Clearly, for equitable outcomes in mathematics for Pāsifika students we need to note the importance of participation as key to learning as Gutiérrez (2002) argues. For too long Pāsifika students' voices have been silenced within traditional mathematics classrooms by the dominant cultural group in Aotearoa/New Zealand.

In this paper, we have described the way in which courageous teachers were willing to shift their pedagogical practices to encompass and build on the cultural beliefs of Pāsifika learners, which were often different from their own.

Their belief in the students is exemplified by a parent's description of the results for their son:

"One day after school, I saw a change in my son's altitude towards maths. He came home after school and instead of throwing down his school bag and asking me what's for dinner, this time he said to me: Dad you know that new Pāsifika Maths? I answered yeah. He said, I like it. I replied, why? He answered back, it's engaging, challenging. I said what does that mean? He answered back, you know talking together about a problem and asking questions. I said good.

The only problem is, he is now using that system to challenge me at home not just in Math's homework but almost anything. But that's okay."

In this statement, the parent has acknowledged the change in his son's mathematical disposition. He has also affirmed that he is culturally comfortable with how his son is using his new voice.

Although there are less than 10% of Pāsifika students currently in schools this group is the fastest growing group of learners in Aotearoa/New Zealand and projected to double within the next decade (Brown et al., 2007; Wylie, 2003). Therefore, the need to address the inequitable situation is urgent and this chapter presents a possible model. In this chapter, we have used a compilation of actions that are representative of how different teachers have drawn on and used Pāsifika values to scaffold their students into the discourse of reasoned mathematical practices promoted by both the RAND group (2003) and Skilling and her colleagues (2016).

This research could be extended to explore the values of other potentially marginalised students from other cultural groups and how teachers can draw on these to increase their participation in the mathematics classrooms but ensuring at the same time, that the students retain their cultural identity.

References

Anae, M., Coxon, E., Mara, D., Wendt-Samu, T., & Finau, C. (2001). *Pāsifika education research guidelines*. Wellington, Ministry of Education.

Ball, D., & Bass, H. (2003). Making mathematics reasonable in school. In J. Kilpatrick, J. Martin, & D. Schifter (Eds.), *A research companion to the principles and standards for school mathematics* (pp. 27–45). Reston, National Council of Teachers of Mathematics.

Boaler, J. (2016). *Mathematical mindsets: Unleashing students' potential through creative math, inspiring messages and innovative teaching.* New York, John Wiley & Sons.

Bourdieu, P., & Passeron, J. C. (1973). *Cultural reproduction and social reproduction.* In R. K. Brown (Ed.), *Knowledge, education and cultural change.* London, Tavistock.

Brown, T., Devine, N., Leslie, E., Paiti, M., Sila'ila'i, E., Umaki, S., & Williams, J. (2007). Reflective engagement in cultural history: A lacanian perspective on Pāsifika teachers in Aotearoa New Zealand. *Pedagogy, Culture & Society, 15*(1), 107–118.

Civil, M., & Hunter, R. (2015). Participation of non-dominant students in argumentation in the mathematics classroom. *Intercultural Journal, 26*(4), 296–312.

Featherstone, H., Crespo, S., Jilk, L., Oslund, J., Parks, A., & Wood, M. (2011). *Smarter together! Collaboration and equity in the elementary math classroom.* Reston, VA, NCTM.

Gutiérrez, R. (2002). Enabling the practice of mathematics teachers in context: Toward a new equity research agenda. *Mathematical Thinking and Learning, 4*(2/3), 145–187.

Hill, J. L., Hunter, J., & Hunter, R. (2019). What do Pasifika students in New Zealand value most for their mathematics learning? In P. Clarkson & W. T. Seah (Eds.), *Values and valuing in mathematics education: Scanning and scoping the territory.* Netherlands, Springer.

Hunter, R. (2008). Facilitating communities of mathematical inquiry. In M. Goos, R. Brown, & K. Makar (Eds.), *Navigating currents and charting directions: Proceedings of the 31st annual conference of the mathematics education research group of Australasia (Vol. 1*, pp. 31–39). Brisbane, MERGA.

Hunter, R. (2013). Developing equitable opportunities for Pasifika students to engage in mathematical practices. In A. M. Lindmeier & A. Heinze (Eds.), *Proceedings of the 37th international group for the psychology of mathematics education (Vol. 3.* pp. 397–406). Kiel, Germany, PME.

Hunter, R., & Hunter, J. (2017). Maintaining a cultural identity while constructing a mathematical disposition as a Pāsifika learner. In E. A. McKinley & L. Tuhiwai Smith (Eds.), *Handbook of indigenous education.* Netherlands, Springer. Crossref DOI link: https://doi.org/10.1007/978-981-10-1839-8_14-1.

Hunter, R., Hunter, J., & Bills, T. (2019). Enacting culturally responsive or socially-response-able mathematics education. In C. Nicol, S. Dawson, J. Archibald, & F. Glanfield (Eds.), *Living culturally responsive mathematics curriculum and pedagogy: Making a difference with/in indigenous communities.* Netherlands, Springer.

Hunter, R., Hunter, J., Anthony, G., & McChesney, K. (2018). Developing mathematical inquiry communities: Enacting culturally responsive, culturally sustaining, ambitious mathematics teaching. *Set, 2,* 25–32.

Hunter, J., Hunter, R., Bills, T., Cheung, I., Hannant, B., Kritesh, K., & Lachaiya, R. (2016). Developing equity for Pāsifika learners within a New Zealand context: Attending to the culture and values. *New Zealand Journal of Educational Studies, 51,* 197–209.

Kazemi, E., Franke, M., & Lampert, M. (2009). Developing pedagogies in teacher educa-tion to support novice teachers' ability to enact ambitious instruction. In R. Hunter, B. Bicknell, & T. Burgess (Eds.), *Crossing divides: Proceedings of the 32nd annual conference of the mathematics education research group of Australasia* (pp. 11–29). Wellington, MERGA.

Paris, D. (2012). Culturally sustaining pedagogy: A needed change in stance, terminology, and practice. *Educational Researcher, 41*(3), 93–97. DOI: 10.3102/0013189X12441244.

RAND Mathematics Study Panel. (2003). *Mathematical proficiency for all students: Towards a strategic research and development program in mathematics education.* Santa Monica, CA, RAND.

Rubie-Davies, C. (2016). High and low expectation teachers: The importance of the teacher factor. In S. T. P. Babel (Ed.). *Interpersonal and intrapersonal expectancies* (pp. 145–157). Abington, Oxon, Routledge.

Smith, M. S., & Stein, M. K. (2011). *5 practices for orchestrating productive mathematics discussion.* Reston, VA, NCTM.

Skilling, K., Bobis, J., Martin, A. J., Anderson, J., & Way, J. (2016). What secondary teachers think and do about student engagement in mathematics. *Mathematics Education Research Journal, 28*(4), 545–566. https://doi.org/10.1007/s13394-016-0179-x

Wylie, C. (2003). *Status of educational research in New Zealand: New Zealand country report.* Wellington, New Zealand Council for Educational Research. Accessed 17 October, 2009, from http://www.nzcer.org.nz/pdfs/12742

The importance of positive classroom relationships for diverse students' well-being in mathematics education

Julia Hill, Margaret L Kern, Jan van Driel and Wee Tiong Seah

Background

Mathematics education is known as a gatekeeper subject, as numeracy skills are necessary for the modern world, impacting upon employment, socio-economic status, health literacy, physical and mental health, and even life expectancy (Plunk, Tate, Bierut, & Grucza, 2014). While a general level of mathematical skills is necessary for everyone, there is additional benefit for those who pursue additional training, with advanced mathematics underlying the data analyses, modelling, forecasting, decision-making, design and management of technological principles that underpin almost all facets of modern enterprise (Statistical Society of Australia, 2005).

Yet in Australia, like in other countries, many students persistently are disengaged from mathematics and harbour negative attitudes and feelings towards mathematics, struggling to complete even basic levels (Clarkson, Bishop, & Seah, 2010). Australian students' performance in international comparative assessments (e.g., Trends in International Mathematics and Science Study [TIMSS]) has not improved in the last 15 years and has in fact dropped to the international average in the 2018 Programme for International Student Assessment (PISA) (Thomson, De Bortoli, Underwood, & Schmid, 2019). Further, the proportion of students studying advanced mathematics in upper high school, and those with the mathematical proficiency to study science, technology, engineering and mathematical (STEM) courses at university has declined over the past thirty years (Kennedy, Lyons, & Quinn, 2014).

Despite the educational, occupational and life benefits of mathematics education, many students, especially at the secondary level, perceive mathematics to be boring, unenjoyable, dull, irrelevant and inaccessible (Attard, 2013). From the middle school years, many students develop 'mathematics anxiety', or a fear of doing mathematics (Geist, 2010). Students then perform poorly on coursework and standardised tests, further fuelling anxieties about the subject, in what can become a vicious negative cycle. The debilitating effects of mathematics anxiety across the lifespan have been well documented (e.g., Dowker, Sarkar, & Looi, 2016; Geist, 2010). For instance, mathematically anxious students are less likely to study mathematics beyond secondary school and pursue careers that require mathematics (Dowker, Sarkar, & Looi, 2016).

These trends point to poor student well-being in many mathematics classrooms. We use the word well-being to refer to how a student subjectively feels and functions. Students with high well-being are engaged with their learning, generally feel good, and are able to manage the difficulties that a subject might bring. Students with low well-being are more likely to disengage, experience anxiety and struggle with their learning. Skills that support well-being in mathematics are often overlooked, with a primary focus on student (under)achievement that emphasises the academic failures of students (Organisation for Economic Cooperation and Development [OECD], 2019; Thomson, De Bortoli, Underwood, & Schmid, 2019). Yet well-being has been found to support student learning outcomes as effectively as cognitive skills (OECD, 2015).

Notably, student well-being is increasingly becoming an important priority for organisations, government and schools around the world. For instance, the OECD, which includes 37 developed and developing countries, proposed a 2030 Learning Compass, which targets student agency, values and well-being as key developmental outcomes for school education (OECD, 2020). The United Nations (2020) includes well-being as one of their 17 goals for sustainable development to achieve by 2030. The COVID-19 pandemic has further highlighted the importance of well-being, with many schools and communities placing the support of mental health above academic performance.

The goals, priorities, policies and initiatives that have been developed on student well-being are important for the holistic development of our young people, helping students to become empowered, contributing members of society. Still, these efforts have focused on global aspects of well-being, assuming that a general, overall indication of functioning is sufficient, similarly to how a student's overall grade point average (GPA) provides a general indication of academic functioning.

Yet generalised scores can obscure important individual variations. Two students with the same GPA can present very different profiles, with one excelling in maths and failing in literacy, another excelling in arts while struggling in science. Educators look to differentiated scores to best adjust learning to the specific learner's needs. Similarly, well-being can vary across different subjects, but few studies have examined how well-being appears within specific subjects. This chapter aims to build knowledge on student well-being specific to mathematics education.

General student well-being

The concept of well-being – also referred to as 'happiness', 'thriving', 'flourishing' and other terms – has many uses and conceptualisations across a broad range of disciplines. In this chapter, we focus specifically on subjective well-being, or a person's subjective perceptions and experiences of feeling and functioning well across a number of dimensions (e.g., physical, mental, social, cognitive). For example, Seligman (2011) proposed that flourishing comprises five well-being dimensions, positive emotions, engagement, relationships, meaning and accomplishment (PERMA). Moore and Lippman (2006) suggested that child and adolescent well-being includes life satisfaction generosity, hope, connectedness, spirituality, self-regulation and prosocial orientations. Kern, Benson, Steinberg, and Steinberg (2016) included five dimensions of adolescent well-being: engagement, perseverance, optimism, connectedness and happiness (EPOCH).

Conceptualising 'mathematical well-being'

An individual's experiences of well-being depend on one's values and may differ across contexts (Alexandrova, 2017; Kern et al., 2020). For instance, a person that values face-to-face social connection might be strongly impacted by the isolation of the COVID-19 lockdowns, whereas a person with social anxiety who values privacy might benefit from the same policies.

Applied to a school context, a student's well-being likely differs depending on the extent to which they value different classes, as well as the underlying values within each subject. As such, there is a need to consider well-being specifically within different subjects. As summarised in Table 7.1, several well-being dimensions have been linked to various positive mathematics learning outcomes.

For instance, positive classroom relationships (e.g., supportive teacher-student relationships) have been associated with improved mathematical achievement, engagement, self-efficacy and interest in mathematics (Battey, 2013; Hackenberg, 2010; Hattie, 2008; Riconscente, 2014; Sakiz, Pape, & Hoy, 2012). Thus, dimensions from current well-being models may also apply to models of mathematical well-being.

To our knowledge, only two studies have specifically explored 'mathematical well-being' (Clarkson, Bishop, & Seah, 2010; Part, 2011). Clarkson, Bishop and Seah proposed a framework of 'mathematical well-being' (MWB) based on Bloom's taxonomic development, which ranged from stage one (low MWB) to stage five (high MWB). According to Clarkson, Bishop, & Seah (2010), high MWB was achieved through development in three domains: cognitive (mathematical skills/knowledge), affective (or values) and emotional (feelings towards mathematics). The authors proposed that high MWB stimulates students' confidence, engagement and positive attitudes in mathematics.

Part (2011) explored mathematical well-being in adult learners, proposing a two-dimensional framework in terms of an individual's capabilities and functioning. Capabilities refer to the things an individual may value doing or

Table 7.1 Dimensions of general well-being models applied to mathematic education with links to positive mathematics learning outcomes (adapted from Hill, Kern, Seah, & van Driel, in press)

Well-being dimension	Description and application to mathematics education	Example of studies finding benefit for mathematics learning
Accomplishment	A sense of achievement, reaching goals or mastery completing mathematical tasks and tests	Keys, Conley, Duncan, and Domina (2012); OECD (2019)
Cognition	A sense of having the knowledge, skills and understanding that is required to do mathematics at school	Kilpatrick, Swafford, and Findell (2001); McPhan et al. (2008); Montague and Van Garderen (2003)
Engagement	A sense of concentration, absorption, deep interest or focus when learning/doing mathematics	Attard (2013); Fielding-Wells and Makar (2008); Høgheim and Reber (2015)
Meaning	Having a sense of direction in mathematics, feeling mathematics is valuable, worthwhile or has a purpose	Gaspard et al. (2015); Hill (2018); Priniski, Hecht, and Harackiewicz (2018)
Perseverance	A sense of drive, or grit, or working hard towards completing a mathematical task or goal	Bass and Ball (2015); Sengupta-Irving and Agarwal (2017); Sullivan et al. (2013)
Positive emotions	Positive emotions when learning/doing mathematics, such as enjoyment, optimism, fun and happiness	Pinxten et al. (2014); Sakiz, Pape, & Hoy (2012); Villavicencio and Bernardo (2016)
Relationships	Having supportive relationships with others, believing one is valued and cared for, connected with others, or supporting peers in mathematics	Battey (2013); Goos (2004); Hattie (2008); Hunter (2008)

being in mathematics, and functioning include valued outcomes in mathematics, such as participating in mathematical discussions or feeling respected in class. Part (2011) suggests that high mathematical well-being entails students feeling that they are capable of being an ideal mathematics student (high capability), whilst also believing they possess the skills required to do mathematics (high functioning).

While these models both use the term 'mathematical well-being', they ignore the important social aspects of mathematics learning. Both lack corresponding measures and they have not been applied to real school settings. The various models described above arise from theory, rather than from the voices and experiences of students themselves. Further, the models are based on primarily Western, developed backgrounds, which may differ across cultures, due to the differing values that different cultures hold.

There is a need to not only better understand mathematics well-being from students' perspectives, but also identify how specific dimensions might transcend specific cultures, in a manner that can be useful to both developed and developing nations. To address these issues, we drew on responses of 488 culturally diverse Year 8 students in Australia to a survey. We aimed to identify factors – in students' own words – that enable them to thrive in mathematics education. We explored differences across ethnic backgrounds, identifying both commonalities and distinctions that arise amongst different sub-cultures.

Our study

Participants included 488 Year 8 students (255 males; 233 females) aged between 13 and 14 years, from nine urban and regional secondary schools (three private; four government/public; two Catholic) in Melbourne and surrounding cities in Victoria, Australia. Schools serviced socio-economic communities ranging from low to high. Students self-identified their ethnicities as Australian (71%), Asian (14%), European (6%), Indian/Pakistani (6%), Indigenous Australian (2%), Middle Eastern (1%) or North/South American (2%). Eighty percent of students were Australian born and 20% were born overseas. For students born overseas, the time spent in Australia ranged from 1 to 14 years.

We focused on students' qualitative responses to the free answer survey question: What makes you feel really good and function well in maths? Students answered this question as part of an online survey.

We imported the responses into NVivo (Version 11) and analysed the responses using a combined deductive/inductive thematic approach adapted from Braun and Clarke (2006). The inductive approach involved scanning the data and generating nodes. For example, 'having friends in my class to support me' was coded as peer support. Single student responses could be coded into multiple nodes. For example, 'I function well when I work in groups, and the class is interesting' was coded both as peer support and interesting/hands on maths.

The deductive approach involved organising and making sense of nodes based on the seven well-being dimensions summarised in Table 7.1 (Hill, Kern, Seah, & van Driel, in press). Frequency counts for each node/dimension were then imported in SPSS (version 25), and then responses were coded as 0 if the node/dimension was not mentioned, or as 1 if the node/dimension was mentioned, resulting in 633 individual nodes classified under eight dimensions.

Table 7.2 summaries the 21 unique nodes that emerged, along with sample student responses and related well-being dimensions. Across all students, positive classroom relationships was by far the most common dimension, followed by a sense of engagement, mathematical understandings and cognitive accomplishments, positive emotions, perseverance, music and meaningful mathematics leaning. Music was mentioned several times with numerous students suggesting listening to music facilitated engagement and

Table 7.2 Classification of student responses (full sample, 633 nodes, N = 488)

Dimension/Nodes	Student examples	Count	%
Relationships		200	31.6%
Teacher support	A supportive or good teacher	94	14.8%
Peer support	Having friends to help me	83	13.1%
General support	When I get help with my learning	38	6.0%
Engagement		126	19.9%
Interesting/hands on	Learning interesting stuff	55	8.7%
Focused working	Being absorbed in my work	37	5.8%
Independent/quietness	When it is quiet and I'm by myself	27	4.3%
Music (engagement)	Music helps me concentrate well	15	2.4%
Cognitive	When I understand the material	96	15.2%
Accomplishment		86	13.6%
Good marks	When I do good in a test	31	4.9%
Accuracy	When I get the answers right	24	3.8%
General mastery	When successful at learning something	17	2.7%
Completing tasks	When I complete my work	13	2.1%
Confidence	When I'm really confident	3	0.5%
Positive emotions		60	9.5%
Enjoyment/fun/happy	If the maths class is enjoyable	47	7.4%
Relaxed/no pressure	When there is no pressure	12	1.9%
Music (emotions)	Music to listen to, to enjoy it more	4	0.6%
Perseverance		31	4.9%
Challenge	Having work I find challenging	21	3.3%
Working hard/practice	When I work hard	13	2.1%
Music (no reasoning)	Listening to music in class	24	3.8%
Meaning		10	1.6%
Future skills	Knowing these skills will help me in life	5	0.8%
Real-world relevance	I like it when problems relate to real life	5	0.8%

positive mood in mathematics. As such, music appears to contribute to the other dimensions, rather than being a separate dimension of well-being. Yet 3.8% (*n* = 24) of students provided no additional reasoning why music was a factor of their well-being and it is unclear the role music plays for each of them. As such these responses appear separately.

Table 7.3 identifies the relative frequencies for the main dimensions across the different ethnicities. At the broad dimensional level, that there were no clear differences across the five ethnicities, nor were there differences for foreign versus Australian born. However, specific patterns differed across the groups.

Whereas relationships, engagement and accomplishment were consistently mentioned across ethnic groups, other dimensions varied by ethnicity. For instance, cognition was commonly mentioned by students with an Australian, Asian and Indian/Pakistani background, with only a few mentions by students with a European background and no mention by students with an Indigenous background.

Table 7.3 Percentage of students from each ethnicity/birth country mentioning each well-being dimension

Dimensions	Aust (n = 345)	Euro (n = 27)	Asian (n = 69)	Indig (n = 8)	Ind/Pak (n = 29)	Aust born (n = 386)	OS born (n = 98)
Relationships	43%	37%	32%	38%	48%	43%	33%
Engagement	28%	19%	23%	38%	17%	27%	22%
Cognition	20%	7%	20%	0%	17%	20%	17%
Accomplishment	16%	19%	28%	13%	17%	16%	25%
Positive emotions	14%	15%	6%	13%	7%	13%	8%
Perseverance	6%	7%	12%	0%	7%	6%	8%
Music	6%	4%	3%	0%	0%	5%	4%
Meaning	3%	0%	1%	0%	0%	2%	1%

Note. Columns can sum to more than 100% as students often mentioned multiple dimensions.

Positive emotions was commonly mentioned for Australians, Europeans and Indigenous, and less relevant to Asian and Indian/Pakistani. Perseverance was relatively more frequent for students with Asian ethnicities. Meaning and Music were consistently only mentioned by a few students. This suggests that at a broad level, there are some dimensions that may be more universal, and others that are more specific to the values of that culture.

Further exploring the relationships dimension

Further insights into similarities and differences across cultures arise through how the broad dimensions are expressed. As relationships was consistently the most referenced dimension across all ethnic groups, we further explored how that dimension was expressed, based on further categorising nodes into sub-nodes and specific ways in which these sub-nodes were described across ethnicities. As summarised in Table 7.4, the relationship dimension included three nodes (teacher support, peer support and general support), which further broke into 10 sub-nodes. While responses did not substantially differ across ethnicities, differences still arose in some of the specific patterns and responses within the sub-nodes.

Teacher support was the most common node for all ethnicities except European students, who focused more on peer support. Students noted the importance of a 'good' or supportive teacher, for example, 'Having a good teacher' (Australian student) or 'Having a really good and supportive teacher' (Asian). Clear explanations from the teacher were frequently cited, with one Indian/Pakistani student noting 'When my teacher gives a great explanation'.

Several students also noted the importance of a helpful teacher, including personalised or one-on-one support ('A teacher who helps each student individually when they struggle', Australian). Students frequently referenced the importance of engaging/fun teachers ('Having a fun maths teacher'). Several students also noted a teacher who understood when individual students

Table 7.4 Percentage of students across each ethnicity mentioning each node and sub-node

Nodes/Sub-nodes	Aust (n = 345)	Euro (= 27)	Asian (n = 69)	Indig (n = 8)	Ind/Pak (n = 29)
Teacher support	18%	15%	20%	22%	34%
Good teacher	6%	7%	7%	11%	10%
Clear explanations	4%	4%	6%	0%	14%
Helpful/1-on-1	5%	0%	2%	0%	4%
Fun/engaging	2%	4%	3%	0%	3%
Understands student	1%	0%	0%	11%	0%
Mood/no pressure	1%	0%	4%	0%	0%
Peer support	18%	22%	9%	13%	21%
Friendships	10%	7%	4%	0%	14%
Sharing/collaboration	8%	19%	3%	13%	10%
Not distracting	2%	4%	0%	0%	0%
General support	9%	0%	6%	0%	6%
Help from anyone	8%	0%	4%	0%	3%
Respectful/valued	1%	0%	1%	0%	3%

Note. Single student responses could be coded into multiple nodes and sub-nodes.

required assistance ('Asking me if understand everything', Indigenous Australian), or teachers with a positive mood rather than pressuring students 'When I have a positive and enthusiastic teacher' (Australian); 'When the teacher is not pressuring me' (Asian).

Peer support was also referenced by students across all ethnicities, especially for Australian and European students. Peer support included references to friendships in the mathematics class or the support of specific friends, with one Australian student noting 'Sitting with my friends'. Other students noted the benefits of sharing ideas, collaboration and group work, for example, 'When I can talk through problems and find solutions with my peers' (European). A smaller proportion of students also noted that working with specific peers was important and facilitated their engagement whilst certain students were distracting ('Sitting next to someone I work well with, someone who isn't distracting', Australian).

General support was mentioned by Australian, Asian and Indian/Pakistani students, but not by European or Indigenous students. Notably, whereas most ethnicities spread across sub-nodes, Indigenous student responses (*n* = 8) clustered into three specific categories: good teacher, understands student and sharing/collaboration.

Findings in perspective

Drawing on the responses of 488 culturally diverse students in Australia, we examined conceptions of well-being from the students' perspectives. We identified both commonalities and distinctive patterns, based on ethnic background, pointing to the way that one's values and background can impact upon one's well-being experiences within mathematics. Notably, relationships were most important, regardless of ethnic background.

On the one hand, this might be surprising, considering Australia tends towards being a more individualistic culture, valuing individual over collective goals and learning and work motivation driven by personal accomplishment and competition (Hofstede, 2011). After all, studies suggest that relationships would be more important for collectivist cultures, such as Indigenous and Asian ethnicities (e.g., Averall, 2012; Hunter et al., 2016).

One explanation may be that the students who self-identified as Australian may value relationships to support individualistic purposes (e.g., higher grades) more than the reciprocal aspects (e.g., helping their peers). For instance, for these Australian students, peer support may be valued because sharing ideas with peers facilitates better mathematical understanding and subsequently higher grades on tests (aligned with individualist values), rather than valuing peer support to improve the mathematical understanding of their peers (aligned with collectivist values). On the other hand, the well-being literature clearly points to relationships being critical to well-being, regardless of culture (e.g., Allen et al., 2017; Baumeister et al., 1995; Ryan & Deci, 2001).

Students referenced a 'good' teacher as one who is engaging or fun, who makes mathematical explanations with clarity, a teacher who understands the learning needs of his/her students and also provides individualised assistance when required. The importance of teachers for students' mathematical well-being is no surprise, considering the central role teachers play for a plethora of positive mathematical outcomes including mathematical engagement, effort, mathematical beliefs, achievement, positive emotions, enjoyment and self-efficacy (Attard, 2013; Grootenboer & Marshman, 2015; Hattie, 2008; Sakiz, Pape, & Hoy, 2012).

For instance, research points to teachers as the most significant factor for student mathematical achievement (Hattie, 2008; Winheller, Hattie, & Brown, 2013), and students' beliefs, attitudes, values and feelings about mathematics are often most impacted by students' perceptions of and relationship with their mathematics teachers (Attard, 2013; Grootenboer & Marshman, 2015; Hill, 2017; Riconscente, 2014).

Studies point to student perceptions of a 'good' mathematics teacher or lesson to involve clear, systematic and detailed explanations from the mathematics teachers (Anthony, 2013; Hill, 2017; Österling, Grundén, & Andersson, 2015; Seah & Peng, 2012).

Conversely, studies find that poor mathematics teaching practices and learning outcomes are often associated with negative relationships between students and their teachers and particularly for students from minority or impoverished backgrounds (Averall, 2012; Battey, Neal, Leyva, & Adams-Wiggins, 2016; Hunter et al., 2016; Jerome, Hamre, & Pianta, 2009; Pianta & Stuhlman, 2004).

Peer relationships were cited almost as frequently as teacher support. Notably, Australian and European students referenced peer support equal to or more than teacher support. Students' referenced the role of friendships, sharing ideas and group work as components of their mathematical well-being. Peer collaboration can support students' problem solving abilities (Fawcett & Garton, 2005), improve mathematical achievement (Whicker, Bol, & Nunnery, 1997) and facilitate positive self-concept and attitudes towards mathematics (Tsuei, 2012). Many students report that they enjoy working with their peers in mathematics (Whicker, Bol, & Nunnery, 1997).

The importance of peer support (along with teacher support) is well documented amongst minority students from collectivist cultures (e.g., Bills & Hunter, 2015; Hill, 2017; Hunter & Anthony, 2011). Yet surprisingly, in our study, peer support was mentioned less frequently than teacher support by the minority students.

This may be explained in part by the fact that the ethnic minority cultures from the top five ethnicities represented in this study exhibit high power distance (Hofstede, 2011) and are those in which teachers are revered (such as the ethnic Chinese with its Confucian Heritage Culture). Notably, despite the many benefits of peer support across all cultural groups, many mathematics classrooms limit teacher-centred instruction and opportunities for peer collaboration (Geist, 2010). The prevalence that students noted the importance of positive relationships for helping them feel and function well suggests that greater attention to the social aspects of mathematics learning might help support students' mathematical well-being, and well-being more generally.

Conclusion

Mathematics empowers individuals acting as a gatekeeper to better health outcomes, employment and full participation in society (Plunk, Tate, Bierut, & Grucza, 2014). Thus, developing mathematical proficiency is paramount for the well-being and prosperity of both individuals' and nations. Yet mathematics education is challenged by persistent student disengagement and poor attitudes towards the subject (Clarkson, Seah, & Pang, 2019). Mathematics education is often pre-occupied with developing students' cognitive skills and raising academic achievement, whilst discounting the importance of student well-being.

Despite a growing international focus on student well-being in schools, student well-being in mathematics education remains poorly understood. Our study revealed what students from different ethnic backgrounds living within Australia consider to be important for their mathematical well-being.

At a broad level, we find more consistencies than differences across ethnicities, suggesting that the general well-being dimensions identified in developed countries may translate well to developing countries. Still, how these dimensions are manifested does vary, such as students from different backgrounds may seek and benefit from different forms of support.

As a whole, our study suggests that that there is more to mathematics education than merely developing cognitive skills and achieving academic benchmarks. Students' mathematical well-being also matters. The words of students captured here provides some hints as to how to support students well in mathematics education.

References

Alexandrova, A. (2017). *A philosophy for the science of well-being*. Oxford, Oxford University Press.

Allen, K. A., & Kern, M. L. (2017). *School belonging in adolescents: Theory, research and practice*. Singapore: Springer Briefs in Psychology.

Anthony, G. (2013). Student perceptions of the 'good' teacher and 'good' learner in New Zealand classrooms. In G. A. B. Kaur, M. Ohtani, & D. Clarke (Eds.), *Student voice in mathematics classrooms around the world* (pp. 209–225). Rotterdam, Netherlands, Sense Publishers.

Attard, C. (2013). "If I had to pick any subject, it wouldn't be maths": Foundations for engagement with mathematics during the middle years. *Mathematics Education Research Journal, 25*(4), 569–587.

Averall, R. (2012). Reflecting heritage cultures in mathematics learning: The views of teachers and students. *Journal of Urban Mathematics Education, 5*(2), 157–181.

Bass, H., & Ball, D. L. (2015). Beyond "You can Do It!" Developing Mathematical Perseverance in Elementary School. Retrieved from www.spenser.org

Battey, D. (2013). "Good" mathematics teaching for students of color and those in poverty: The importance of relational interactions within instruction. *Educational Studies in Mathematics, 82*(1), 125–144.

Battey, D., Neal, R. A., Leyva, L., & Adams-Wiggins, K. (2016). The interconnectedness of relational and content dimensions of quality instruction: Supportive teacher-student relationships in urban elementary mathematics classrooms. *Journal of Mathematical Behavior, 42*, 1–19.

Baumeister, R. F., & Leary, M. R. (1995). The need to belong: Desire for interpersonal attachments as a fundamental human motivation. *Psychological Bulletin, 117*(3), 497–529.

Bills, T., & Hunter, R. (2015). The role of cultural capital in creating equity for Pāsifika learners in mathematics. In *Proceedings of the 38th annual conference of the Mathematics Education Research Group of Australasia* (pp. 109–116).

Braun, V., & Clarke, V. (2006). Using thematic analysis in psychology. *Qualitative Research in Psychology, 3*(2), 77–101.

Clarkson, P., Bishop, A. J., & Seah, W. T. (2010). Mathematics education and student values: The cultivation of mathematical wellbeing. In T. Lovat, R. Toomey, & N. Clement (Eds.), *International research handbook on values education and student wellbeing* (pp. 111–135). Dordrecht, Springer.

Clarkson, P., Seah, W. T., & Pang, J. (2019). Scanning and scoping of values and valuing in mathematics education. In P. Clarkson, W. T. Seah, & J. Pang (Eds.), *Values and valuing in mathematics education: Scanning and scoping the territory* (pp. 1–10). Cham, Springer.

Dowker, A., Sarkar, A., & Looi, C. Y. (2016). Mathematics anxiety: What have we learned in 60 years? *Frontiers in Psychology, 7*, 508.

Fawcett, L. M., & Garton, A. F. (2005). The effect of peer collaboration on children's problem-solving ability. *British Journal of Educational Psychology, 75*(2), 157–169.

Fielding-Wells, J., & Makar, K. (2008). Student (dis) engagement in mathematics. In P. L. Jeffery (Ed.), *Annual conference of the Australian Association for Research in Education* (pp. 1–10). Coldstream, AARE.

Gaspard, H., Dicke, A.-L., Flunger, B., Brisson, B. M., Häfner, I., Nagengast, B., & Trautwein, U. (2015). Fostering adolescents' value beliefs for mathematics with a relevance intervention in the classroom. *Developmental Psychology, 51*(9), 1226.

Geist, E. (2010). The anti-anxiety curriculum: Combating math anxiety in the classroom. *Journal of Instructional Psychology, 37*(1), 24–31.

Goos, M. (2004). Learning mathematics in a classroom community of inquiry. *Journal for Research in Mathematics Education, 35*(4), 258–291.

Grootenboer, P., & Marshman, M. (2015). *Mathematics, affect and learning: Middle school students' beliefs and attitudes about mathematics education.* London, Springer.

Hackenberg, A. J. (2010). Mathematical caring relations in action. *Journal for Research in Mathematics Education, 41*(3), 236–273.

Hattie, J. (2008). *Visible learning: A synthesis of over 800 meta-analyses relating to achievement.* Chicago, Routledge.

Hill, J. L. (2017). *What do culturally diverse middle school students value for their mathematics learning? (Unpublished Master's thesis).* Massey University, Auckland. Retrieved from http://hdl. handle.net/10179/13356

Hill, J. L. (2018). What do culturally diverse students in New Zealand value for their mathematics learning? In G. Anthony, J. Dindyal, & V. Geiger (Eds.), *Proceedings of the 41st annual conference of the Mathematics Education Research Group of Australasia* (pp. 384–391). Auckland, New Zealand: MERGA.

Hill, J. L., Kern, M. L., Seah, W. T., & van Driel, J. (2020). Feeling good and functioning well in mathematics education: Exploring students' conceptions of mathematical wellbeing and values. *ECNU Review of Education.* https://doi.org/10.1177/2096531120928084

Hofstede, G. (2011). Dimensionalizing cultures: The Hofstede model in context. *Online Readings in Psychology and Culture, 2*(1). https://doi.org/10.9707/2307-0919.1014.

Høgheim, S., & Reber, R. (2015). Supporting interest of middle school students in mathematics through context personalization and example choice. *Contemporary Educational Psychology, 42*, 17–25.

Hunter, J., Hunter, R., Bills, T., Cheung, I., Hannant, B., Kritesh, K., & Lachaiya, R. (2016). Developing equity for Pāsifika learners within a New Zealand context: Attending to culture and values. *New Zealand Journal of Educational Studies, 51*(2), 197–209.

Hunter, R. (2008). Facilitating communities of mathematical inquiry. In *Navigating currents and charting directions. Proceedings of the 31st annual conference of the Mathematics Education Research Group of Australasia* (Vol. 1, pp. 31–39).

Hunter, R., & Anthony, G. (2011). Forging mathematical relationships in inquiry-based classrooms with Pasifika students. *Journal of Urban Mathematics Education, 4*(1), 98–119.

Jerome, E. M., Hamre, B. K., & Pianta, R. C. (2009). Teacher–child relationships from kindergarten to sixth grade: Early childhood predictors of teacher-perceived conflict and closeness. *Social Development, 18*(4), 915–945.

Kennedy, J. P., Lyons, T., & Quinn, F. (2014). The continuing decline of science and mathematics enrolments in Australian high schools. *Teaching Science, 60*(2), 34–46.

Kern, M. L., Benson, L., Steinberg, E. A., & Steinberg, L. (2016). The EPOCH measure of adolescent well-being. *Psychological Assessment, 28*(5), 586.

Kern, M. L., Williams, P., Spong, C., Colla, R., Sharma, K., Downie, A., Taylor, J. A., Sharp, S., Siokou, C., & Oades, L. G. (2020). Systems informed positive psychology. *Journal of Positive Psychology*, 15, 705–715

Keys, T. D., Conley, A. M. M., Duncan, G. J., & Domina, T. (2012). The role of goal orientations for adolescent mathematics achievement. *Contemporary Educational Psychology*, 37(1), 47–54.

Kilpatrick, J., Swafford, J., & Findell, B. (2001). *Adding it up: Helping children learn mathematics*. Washington, National Academy Press.

McPhan, G., Morony, W., Pegg, J., Cooksey, R., & Lynch, T. (2008). *Maths? Why not? Final report*. Canberra, Department of Education, Employment and Workplace Relations (DEEWR).

Montague, M., & Van Garderen, D. (2003). A cross-sectional study of mathematics achievement, estimation skills, and academic self-perception in students of varying ability. *Journal of Learning Disabilities*, 36(5), 437–448.

Moore, K. A., & Lippman, L. H. (2006). *What do children need to flourish?: Conceptualizing and measuring indicators of positive development*. Washington, Springer.

OECD (2015). *Skills for social progress: The power of social and emotional skills*. Paris, France, OECD Publishing.

OECD (2019). *PISA 2018*. Paris, OECD Publishing.

OECD. (2020). OECD Learning Compass 2030. Available at http://www.oecd.org/education/2030-project/teaching-and-learning/learning/learning-compass-2030/. Retrieved May 14, 2020.

Österling, L., Grundén, H., & Andersson, A. (2015). Balancing students' valuing and mathematical values. In S. Mukhopadhyay & B. Greer (Eds.), *Eighth International mathematics education and society Conference* (Vol. 3, pp. 860–872). Porland, OR, MES.

Part, T. (2011). What is 'mathematical well-being'? What are the implications for policy and practice? In C. Smith (Ed.), *Proceedings of the British Society for Research into Learning Mathematics* (Vol. 31, pp. 121–126). London, BSRLM.

Pianta, R. C., & Stuhlman, M. W. (2004). Teacher-child relationships and children's success in the first years of school. *School Psychology Review*, 33(3), 444–458.

Pinxten, M., Marsh, H. W., De Fraine, B., Van Den Noortgate, W., & Van Damme, J. (2014). Enjoying mathematics or feeling competent in mathematics? Reciprocal effects on mathematics achievement and perceived math effort expenditure. *British Journal of Educational Psychology*, 84(1), 152–174.

Plunk, A., Tate, W. F., Bierut, L. J., & Grucza, R. A. (2014). Intended and unintended effects of state-mandated high school science and mathematics course graduation requirements on educational attainment. *Educational Researcher*, 43(5), 230–241.

Priniski, S. J., Hecht, C. A., & Harackiewicz, J. M. (2018). Making learning personally meaningful: A new framework for relevance research. *The Journal of Experimental Education*, 86(1), 11–29.

Riconscente, M. M. (2014). Effects of perceived teacher practices on Latino High school Students' interest, self-efficacy, and achievement in mathematics. *The Journal of Experimental Education*, 82(1), 51–73.

Ryan, R. M., & Deci, E. L. (2001). On happiness and human potentials: A review of research on hedonic and eudaimonic well-being. *Annual Review of Psychology*, 52, 141–186.

Sakiz, G., Pape, S., & Hoy, A. (2012). Does perceived teacher affective support matter for middle school students in mathematics classrooms? *Journal of School Psychology*, 50(2), 235–255.

Seah, W. T., & Peng, A. (2012). What students outside Asia value in effective mathematics lessons: A scoping study. *ZDM: The International Journal on Mathematics Education, 44*(1), 71–82.

Seligman, M. E. P. (2011). *Flourish: A visionary new understanding of happiness and well-being.* Free Press.

Sengupta-Irving, T., & Agarwal, P. (2017). Conceptualizing perseverance in problem solving as collective Enterprise. *Mathematical Thinking and Learning, 19*(2), 115–138.

Statistical Society of Australia. (2005). Statistics at Australian universities: An SSAI-sponsored review. Braddon, ACT.

Sullivan, P., Aulert, A., Lehmann, A., Hislop, B., Shepherd, O., & Stubbs, A. (2013). Classroom culture, challenging mathematical tasks and student persistence. In V. Steinle, L. Ball, & C. Bardini (Eds.), *Paper presented at the annual meeting of the Mathematics Education Research Group of Australasia* (pp. 618–625). Melbourne, Australia, MERGA.

Thomson, S., De Bortoli, L., Underwood, C., & Schmid, M. (2019). *PISA 2018: Reporting Australia's Results. Volume 1 Student Performance.* Retrieved from https://research-acer-edu-au.ezp.lib.unimelb.edu.au/ozpisa/35

Tsuei, M. (2012). Using synchronous peer tutoring system to promote elementary students' learning in mathematics. *Computers and Education, 58*(4), 1171–1182.

United Nations. (2020). Sustainable Development Goals. Available at: www.un.org/sustainabledevelopment/. Retrieved January 10, 2020.

Villavicencio, F. T., & Bernardo, A. B. I. (2016). Beyond math anxiety: Positive emotions predict mathematics achievement, self-regulation, and self-efficacy. *Asia-Pacific Education Researcher, 25*(3), 415–422.

Whicker, K. M., Bol, L., & Nunnery, J. A. (1997). Cooperative learning in the secondary mathematics classroom. *The Journal of Educational Research, 91*(1), 42–48.

Winheller, S., Hattie, J. A., & Brown, G. T. L. (2013). Factors influencing early adolescents' mathematics achievement: High-quality teaching rather than relationships. *Learning Environments Research, 16*(1), 49–69.

Chapter 8

Novel frameworks for upskilling the mathematics education workforce

Padmanabhan Seshaiyer

Introduction

In this work, we will introduce some novel educational frameworks that provide the opportunity for mathematics educators to not only engage students in primary to secondary grades and beyond through mathematical tools to represent, understand and solve real-world problems but also engage them in using tools to make a decision, prediction or solution about a real-world problem. These frameworks also help to prepare students to become life-long learners who can then go on to pursue state-of-the-art jobs. To create this next generation cohort of students, however, we must upskill our current educators with pedagogical practices that go beyond just delivering content. In particular, mathematics education must include a variety of learning approaches including experiential learning, inquiry-based learning, challenge-based learning and interdisciplinary problem-based learning. In this work, we provide some learning frameworks and benchmark example mathematical tasks to compare the effectiveness of these approaches. We will consider authentic mathematical tasks that incorporate a shared collaborative experience with innovative pedagogical practices to advance mathematics teaching and learning in the 21st century.

We live in an era where the new value proposition for our educational system is to build competencies along with the mathematics content that we teach within the school curricula. Some of these competencies include 21st-century skills, namely, communication, collaboration, critical thinking and creativity. Other competencies include data competency, problem-solving competency, cross-cultural competency, information communication and technology (ICT) competency and global competency. The challenge, however, is that our instruction that is often content-focused based on national standards and textbooks, may not necessarily expose the students to these different competencies explicitly.

Competency maybe defined in particular as the ability of someone's readiness to act in response to challenges of a given situation. In addition, *mathematical competency* is an extension of the general definition of competency when the challenges involved in the definition of competence are mathematical.

The majority of our general educational systems all across the globe still practice the teaching philosophy *"Here is the mathematics content, go solve the real-world problem"*. In this work, we propose some methods on how educators can employ some powerful frameworks to reverse this philosophy to *"Here is the real-world problem, let us find the mathematical competency to do it"*.

Combined with this paradigm mind shift in teaching is an additional challenge to upskill the existing teacher workforce in mathematics education. *Upskilling* in education would be the process of training the current educators' new pedagogical practices. This is important because *passionate educators* are far more likely to *produce excited students* who want to continue learning. In today's age, it is not only the content knowledge that matters but also the 21st-century life-long learning skills required to teach today's children (see Figure 8.1). This would imply that educators must not only focus on developing content knowledge for students but also competencies for being able to apply the knowledge to real-world problems. While content-based instruction may focus on a specific discipline or subject, a competency-based education helps the students to gain transferable skills needed for the workforce. This challenge for the need to upskill teachers is not only there for mathematics but for other areas as well (Peters-Burton et al., 2015; Sentance, Dorling, McNicol, & Crick, 2012; Sin, Tsang, Poon, & Lai, 2010). While there are many ways to do *upskilling*, direct partnerships between mathematics educators and teachers have proven to be very successful in implementing effective practices (Seshaiyer & Kappmeyer, 2016).

In this work, we will consider four specific tasks and describe the mathematical competencies motivated by specific learning frameworks that could be used to solve a given real-world problem. Each task involves a specific

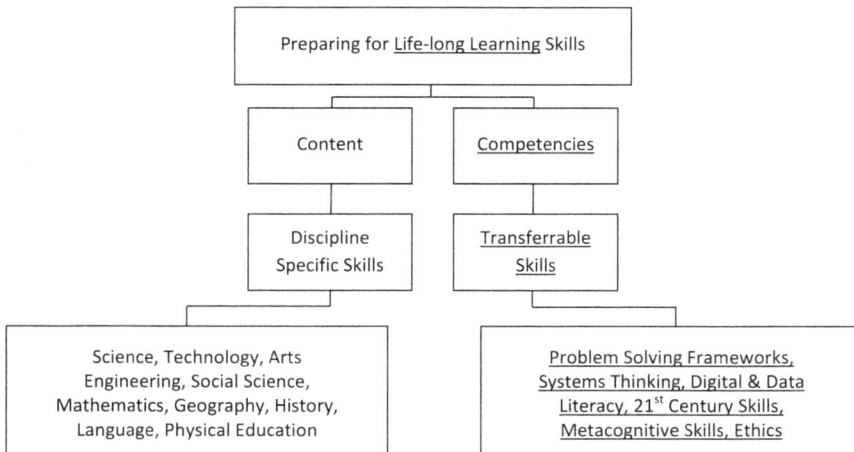

Figure 8.1 Preparing the next-generation towards life-long learning.

educational strategy combined with an educational framework to engage the participants in a mathematical competency. The first task involves a discussion on *experiential learning strategy* combined with an *active learning framework* to build a mathematical competency on *proportional reasoning*. The second task involves a discussion on *inquiry-based learning* motivated through a *5E-instructional framework* to develop mathematical competency of students on *variations*.

The third task involves a discussion on *challenge-based learning* strategy combined with *FERMI problem-solving framework* to develop the mathematical competency on *determining volume*. The fourth and final task involves a discussion on *interdisciplinary problem-based learning* strategy combined with a *modelling* framework to develop the mathematical competency on *finding area*.

These tasks have been tested at multiple professional development workshops given by the author with primary- and secondary-grade teachers as well as in classrooms with students. Specifically in this work, we share some of the discussions from these workshops and classroom observations. These help us as educators to reflect on our own pedagogical practices and how we can adapt to create better 21st-century classrooms that motivate mathematical thinking and learning.

The objective of this chapter is to expose mathematics educators to use the learning frameworks and the mathematical tasks to not only engage the students to improve their mathematical competency but also upskill their own pedagogical practices to prepare students for life-long learning skills.

Developing mathematical competency through *experiential learning*

Students learn best through experience. One of the powerful ways to engage students in learning a subject matter is through an *experiential Learning* strategy (Kolb, 2014). While the phrase "*experiential learning*" is not easy to define, it may be thought as a particular form of learning from life experience that includes spirals of learning with continuous inquiry into the nature of experience and the process of learning from it. This strategy is often contrasted with lecture and classroom learning. As Keeton and Tate (1978) proposed experiential learning to be, "*Learning in which the learner is directly in touch with the realities being studied. It is contrasted with the learner who only reads about, hears about, talks about, or writes about these realities but never comes into contact with them as part of the learning process*". Experiential learning, therefore, emphasizes explicit experiences and actions in response to a prescribed context as the primary source of learning and helping one to discover a role for mathematical competency through mathematical thinking, analysis and content knowledge.

The discussion in this section includes observations from one of the professional development workshops that the author led for primary and secondary mathematics teachers where the following task was presented. A picture of three people standing in a line and looking towards a group of trees, each person holding up a bamboo skewer in their right hand, was shown to all participants and they were asked to engage in an active discussion on what they observed. Specifically, the *active learning* framework involved engaging the participants in *noticing* and *wondering* from the picture (see Rumack & Huinker, 2019, and references therein).

The *notice* responses from the teachers ranged from "*they are children playing in a park*" to "*they are playing roles in Harry Potter with wands*" and so on. Not a single response from any teacher participant included that "*the students are in the back of their school in the playground*". When this fact was disclosed, it was a surprise to most teachers and they curiously asked "*What are the students doing there?*" Some of them even went on to comment, "*the students must be playing outside during recess*". This led the author as an educator to take the teachers from the *notice* phase to the *wonder* phase by asking them "*What do you wonder about the picture now?*"

The *wonder* responses were even more interesting and included everything from "*students trying to find the direction of the wind*" to "*students trying to play a Harry Potter mystery wand game*". At this point, the author gave them a clue that the "*students are in the back of their school in the playground trying to solve a mathematics problem*". This new information made some teacher *wonder* that maybe the *students were trying to find the height of a tree*. Bingo! At that moment, the teachers were told that each of the students were holding a bamboo skewer in their hand. Then all the teacher participants were asked to share their mathematical thinking on how the students would accomplish the task of finding the *height of the tree with nothing more than the bamboo skewer.*

To help them express their ideas, the participants were asked to work in groups and display their solutions from the known facts on a collaborative space with writable classroom walls. Meyers and Jones (1993) had defined such practices involving *learning environments* allowing students to engage in problem-solving exercises in informal small groups and allowing them to communicate, collaborate and reflect as *active learning*. This way of engagement provides active learning through collaboration and makes the classroom more student-centred. There is also evidence that such open collaborative classrooms increase the performance of students (Freeman et al., 2014).

For finding the *height of the tree*, most secondary teacher participants quickly expressed their solutions on the board using similar triangles *geometry* or right triangle *trigonometry*. Some of them wanted to find the angle of elevation first so they can then use trigonometric ratios to find the height of the tree. Others assumed that there is enough sunlight and therefore the students will take advantage of the shadows and similar triangles. Figure 8.2 shows two of the illustrations that teachers shared on the whiteboard.

While both these approaches will help identify equations relating heights to lengths of shadows to calculate the height of the tree, the pictures drawn

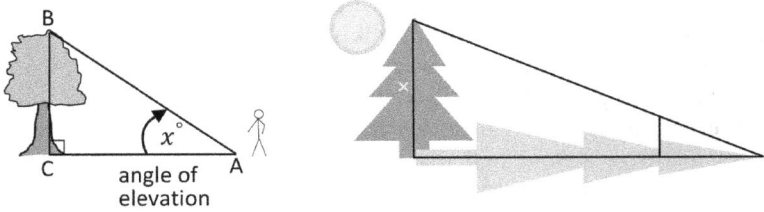

Figure 8.2 An illustration of pictures produced by secondary-grade teachers.

by most of these teachers reminded everyone of standard textbook examples used in a trigonometry section or pictures of sun and shadows with similar triangles. Yes, there may have been some *critical thinking* here but less *creativity* in the solution process. When asked deeper questions such as how one finds the angle of elevation, the teachers indicated that the students could use a surveying instrument for angle measurement. In addition, the teachers were challenged to think about alternate ways to solve if these instruments were not available or if there was *no sunlight* and therefore, *no shadow*.

At this point, these teachers were informed that the students are in third and fourth grade (about 8–9 years old) and they had not been exposed to content such as *similar triangles* or *trigonometry*. What the author was doing at this point was playing devil's advocate to engage the teachers in deeper mathematical thinking. However, many of these teachers could not think any further as they had decided that the way to solve the problem is through *similar triangles* or *trigonometry*. They struggled to think of simpler ways and these teachers started to realize that they were trying to "fit" the mathematics they were trained in to the problem at hand, which was finding the *height of the tree*.

Next, the teachers were given some additional information. What they did not see in the picture is that each of the three students was working in pairs and so they each had a partner who is not seen in the picture but standing between them and the tree. The teachers were given the opportunity next to discuss how they would use this new information to come up with a simple approach that can convince a third- or fourth-grade student who is not yet exposed to higher level mathematics concepts in geometry and trigonometry. While most secondary teachers grappled with this, a group of primary-grade teachers proposed a solution, which was simple, elegant and efficient:

> The student holding the bamboo skewer should walk back far enough so they use the full bamboo skewer to capture the entire tree, they capture the height of their partner on the bamboo skewer, and mark it with a pencil. Now each pair knows the true height of their friend, they know the height of their friend as recorded on the bamboo skewer, they also know the height of the bamboo skewer and that should be enough to set

up a proportional reasoning argument to estimate the height of the tree. With three pairs, one has three estimates. Taking the average of the three can give a reasonable estimate of the height of the tree.

Following this explanation, the primary-grade teachers in pairs demonstrated the process using pencils (instead of a bamboo skewer) and explaining the ratios of heights they had written on the wall to determine the height of the tree. Essentially, they had expressed their work using proportions in two different ways along with a picture next to it that reflected their live *acting it out* demonstration (see Figure 8.3).

Height of tree (x): Length of skewer (pencil) = Height of friend: Length of friend (on skewer(pencil))

$$\frac{Height \ of \ tree \ (x)}{Height \ of \ friend} = \frac{Length \ of \ skewer \ (pencil)}{Length \ of \ friend \ (on \ skewer \ (pencil))}$$

The secondary-grade teachers were awed and surprised to hear this simple and elegant solution along with the explanation through role-playing and noted the related illustrations. They commended the primary-grade teachers in the workshop for the novelty and simplicity of their approach. They saw the primary-grade teachers used mathematical competency including *proportional reasoning, ratios and averages* to solve a real-world problem. The secondary-grade teachers also noticed how eloquently the primary-grade teachers communicated their solution ideas that can be easily relatable by more practitioners and educators. This opportunity provided all the teacher participants to realize the importance of *"finding the mathematical competency needed to solve the problem"* rather than *"fitting the mathematics they know and think will solve the problem"*. It also provided the teachers an opportunity to reflect on how it is important not to always gravitate towards textbook approaches to solve problems. Rather they understood the importance of using experiential learning as an opportunity to enhance the mathematical competency of students.

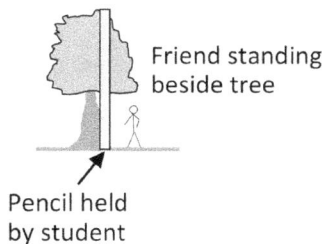

Figure 8.3 An example of explanation produced by primary-grade teachers.

Engaging students in mathematical competency via *inquiry-based learning*

Inquiry-based learning has been a successful educational strategy employed in science education in which students follow methods and practices that help them to construct knowledge (Keselman, 2003; Pedaste et al., 2015). The approach engages students in a process of discovering new cause-effect relations, with the students trying to formulate hypotheses and test them by conducting suitable experiments or making observations (Pedaste, Mäeots, Leijen, & Sarapuu, 2012). Because of the nature of the process, inquiry-based learning has often been associated with science (Gormally et al., 2009). But *what will inquiry-based learning look like to motivate mathematical competency?*

In one of the summer camps hosted by the author, a group of middle school students (ages 10–13 years old) were asked to come into a physics lab and engage with some equipment that was on a table. The equipment consisted of an *open box* inside which, there was a small *open bucket* that held a *solar pump* immersed in water. One end of the pump was connected to a *solar cell* that was under a *lamp* that acted as the sun. The inquiry-based learning objective was to engage the students to discover if the light intensity from the bulb inside the lamp influences the output of the solar cell, and hence affect the release of a fountain from the solar pump. The framework we employed to get the students to learn through an inquiry-based strategy was a 5-E instructional model that consists of five phases including engage, explore, explain, elaborate and evaluate.

We started with the ENGAGE phase where we had the students play with the equipment in groups in their respective tables. This helped to set the purpose of the lesson to drive a goal for an essential topic for the students to engage. Then we let them EXPLORE. Here we gave the students some inquiry-based questions they had to use to identify a hypothesis (preferably mathematical in nature). These questions included:

1 *Currently you have a 40-W bulb in the lamp. If you increase the wattage of the bulb to 60 W, 80 W and then to 100 W, will the height of the fountain go up or down?*
2 *Fixing the bulb at 60 W, if you now half the distance between the solar cell and the lamp, and then further divide the distance again by two and continue this process, will the height of the fountain continue to go up or down?*

There were of course bulbs of different wattage, as mentioned in the question, available for the students to test what they think is going to happen. This EXPLORE phase gives the participants a chance to understand the concepts in depth so they can discover the fundamental topic. The ENGAGE and EXPLORE phases help to build the *collaboration* competency

Table 8.1 Guided inquiry with template for tabular approach

Bulb wattage	Fountain height
40 W	
60 W	
80 W	
100 W	
Distance between Lamp and Solar Cell	Fountain Height
1 feet	
½ feet	
¼ feet	

which helps students to work in teams effectively. To assist them in their inquiry-based approach, the teacher may also provide a table or clues for creating a graph the students can draw to predict the trend. An example is given in Table 8.1.

The next part involved the opportunity for the students in each group to EXPLAIN to students in other groups about their observation. This phase also helps to build the *communication* competency. It also helps the students to develop research and *critical thinking* skills while deepening their understanding of the core topic they are being exposed to. As they do this, the next phase would be to help participants to reflect and refine their thinking in the ELABORATE phase. This may be because they saw other participants explain differently, which made them think more deeply or because their explanation could be improved. At this stage, students can be very *creative* as they find ways to make their hypothesis more concrete. Finally, the students are guided to demonstrate their understanding through an EVALUATE phase that helps them to assess each other's hypothesis and come to a reasonable conclusion.

The reader may have noticed that the mathematical content that was being motivated here is *direct* and *inverse* variations. In particular, the students may note in response to the first question that as the wattage of the bulb increases the height of the fountain also increases. A related graph may help them to understand if this trend is *linear* or *non-linear*. Whatever the trend is they would realize that the trend motivates *direct* variation. Similarly, the response to the second question helps them to understand that the fountain goes higher when the distance between the lamp and the solar cell becomes smaller. In addition, this relation yields an *inverse* variation, which would also look different on a graph. For engaging the students to further their creativity, the teachers may ask if there is a relation between the height of the fountain and the distance between the lamp and the solar cell. This helps them to answer deeper questions such as whether the fountain height is inversely proportional to the distance or the square of the distance or some other power of the distance, which could eventually lead to a potential formula.

Engaging students in mathematics
through *challenge-based learning*

Challenge-based learning is a strategy for enhancing learning while solving real-world challenges. Challenge-based learning combines features from problem-based learning, project-based learning and inquiry-based learning applied to the problems that are authentic and more open-ended. In particular, participants engage in activities that help them to identify big ideas, make meaningful assumptions, ask good questions, gain in-depth subject area knowledge and develop 21st-century skills.

While there has been a growing literature on problem-solving frameworks ever since the introduction of Polya's *How to Solve It* (Polya, 1957), research suggests that the greatest difficulty in the problem-solving process is in the identification of an appropriate mathematical model, which requires contextual knowledge of the real-world situation as well as creativity. One framework to engage students in employing challenge-based learning approach to interpret a real-world context and solve an associate problem is using the *FERMI problem-solving framework* (Ärlebäck & Bergsten, 2013; Peter-Koop, 2004).

Enrico Fermi (1901–1954), who won the Nobel Prize for Physics in 1938 for his work on nuclear processes, was well known for creating open real-world problems that could only be solved by giving a reasonable estimate. Some examples of Fermi problems include, *"How many piano tuners are there in Chicago?"* or *"How many people in the world are talking on their cell phones at this instant?"* These problems are not only very real world but also give the problem solver an indication that the solution will involve big numbers and lot of data. The key in solving these problems involves the need for some additional information and then the type of assumptions that the problem solver would need to make in order to solve the problem.

For example, consider the problem, *"How many piano tuners are there in Chicago?'* The challenge in this problem is to first know something about Chicago. Some related questions students may want to know that would help with the solution strategy are, *'What is the population of Chicago?"*, *"How many families have a piano?"* and *"How many jobs can a piano-tuner do in a day?"* Students that do not know what a piano tuner does, need an explanation of what their actual job is. This is important for teachers to understand as the task automatically assumes that all students know about *Chicago*, job of *piano-tuner*, etc. Suppose we decide to give just one extra item of information to the students, which is that *"the population of Chicago is 3,000,000"*. This will now bring all students in the class to the same starting point. Here is a potential FERMI problem-solving approach that one can follow:

- *Assuming that every family has an average of four people (Dad, Mom and two children), how many families are there in Chicago?*
- *A quick calculation would yield 3,000,000 ÷ 4 = 750,000 families.*

- *Assuming now that every fifth house in Chicago has a piano, how many families have a piano?*
- *A quick calculation would yield 750,000 ÷ 5 = 150,000 pianos.*
- *Let us now try to find how many total jobs one piano-tuner can do. Suppose a piano-tuner can do a maximum of four jobs a day and works for 5 days a week, then the piano-tuner completes 20 jobs a week. If the piano-tuner continues this trend for a year for 50 weeks (assuming they take a vacation for 2 weeks) will then yield a total of 50 × 20 = 1000 jobs completed by one piano-tuner. All that is left to do is to divide the total number of households that have a piano, which is 150,000, by the total number of jobs that one piano-tuner can do which is 1000. This finally, gives the number of piano-tuners to be 150.*

While the FERMI problem-solving framework presented here provides an opportunity to obtain a good estimate to the true answer, the estimate however is based on the assumptions made along the way. Some of these may or may not be correct. For example, there could be a different average number of people in the household based on geography or religion. There could be parts of Chicago where no household owns a piano or regions where every household has a piano. Also note the clever use of 5 working days in a week to make easy calculation and the use of 50 weeks in a year (instead of 52) as that makes the calculation of the estimate much easier. Most importantly, this also helps students to become mathematically competent to deal with data. In particular, students learn to collect data, interpret data, visualize data, predict with data and such data practices helps to enhance their computational thinking skills.

At another professional development workshop with primary-grade teachers, the author asked the following question to motivate challenge-based learning, along with a picture appearing to show a room almost filled with popcorn kernels.

How many popcorn kernels will it take to fill the room you are sitting in? Also, what mathematical competency are we trying to build through this task?

The workshop participants had absolutely no trouble identifying the mathematical competency this activity tried to help build which is *determining volume*. They also recognized that this problem involves big data as they realized the number of popcorn kernels is large but they had to come up with a creative way to determine this number. The teachers were allowed to use any approach they felt would help get a good estimate.

Assuming that every popcorn kernel can fit into a half-inch by half-inch by half-inch cube, one can ask about the number of kernels that can fit into a 1-inch by 1-inch by 1-inch cube. A quick visual analysis indicates that a 1-inch by 1-inch by 1-inch cube contains eight half-inch by half-inch by half-inch cubes and hence contains eight popcorn kernels. Approximating

1-foot as 10 inches instead of 12 inches (for ease of calculation), a 1-foot by 1-foot by 1-foot cube will contain 8000 popcorn kernels. Now that we have a measure for a 1-foot by 1-foot by 1-foot cube, all we have to do is to find an estimate of the length, width and height of the room in feet. That will tell us how many 1-foot by 1-foot by 1-foot cubes can fit into the room with those dimensions.

Note the steps in the estimation that make the problem-solving efficient. For example, assuming 1 foot to be about 10 inches makes the multiplication in the computation easier. Ultimately, one is able to provide an educated guess for the answer to the number of popcorn kernels but also gets a good grasp of the notion of capacity or volume. One can then transition to more challenging exercises involving other three-dimensional solid geometry that can allow students to be more creative.

> Suppose at Theatre A, popcorn is served in a box, at Theatre B, popcorn is served in a cylindrical container and at home, popcorn is served in a bowl (hemispherical in shape). (Figure 8.4)
>
> Disregarding the thickness of the container and any popcorn piled above the container, based upon the given dimensions, at which location are you getting the most popcorn?

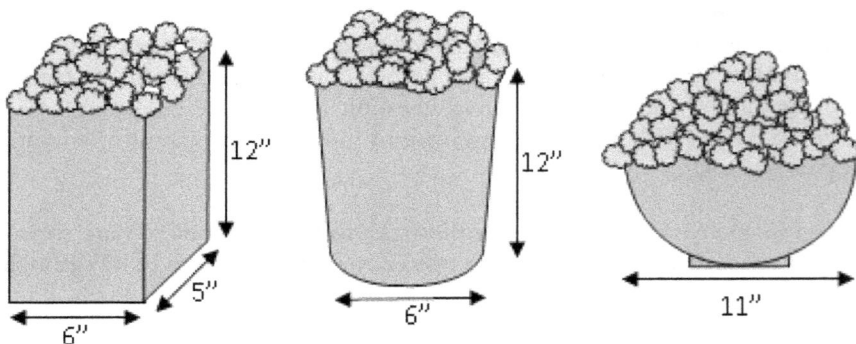

Figure 8.4 Extension and application of FERMI problem-solving to other 3D solids.

Engaging students in mathematical competency through *interdisciplinary PBL*

Problem-based learning (PBL) may be thought of as a way of constructing instruction using problems as stimuli with a focus on student activity (Wood, 2003). While the focus of problem-based learning is not entirely on the exposition of disciplinary content knowledge, the approach helps students to acquire knowledge and skills through enactment of a sequence of steps presented in context with guided support from teachers and learning materials. The approach of problem-based learning becomes more interesting if it is employed outside the silo of a single discipline (Sternberg, 2008). Such problem-based learning approaches can help to teach students the knowledge and skills needed to make them think in an interdisciplinary way. Next, we discuss this approach which may be referred to as interdisciplinary problem-based learning approach and can help build mathematical competency if the problem used as a stimulus comes from a discipline outside of mathematics.

Consider the following problem of finding the area of a bacterial colony inside a petri-dish, one of the fundamental computations that needs to be done in a biology or chemistry lab. See Figure 8.5.

When asked, most teachers in biology or chemistry said that they would use an instrument called the *spectrophotometer* to determine the bacterial density and then the area as an estimate. When the author asked if they would be able to estimate the area without the instrument, most of these teachers said they did not know how to do this, as the shape of the colony was not a standard geometric shape.

This interaction motivated the author to coordinate a professional development workshop with a group of teachers who were teaching biology and chemistry in the secondary grades. At the start of the workshop, the teachers were introduced to the notion of *mathematical modelling* framework as an approach to translate problems from an application area (e.g., biology) into a tractable mathematical formulation whose analysis can help provide important insight or answers that is useful for the original application. The teachers were asked to have a discussion at their respective tables about the subject they teach and identify any specific mathematical modelling techniques they use in their instruction.

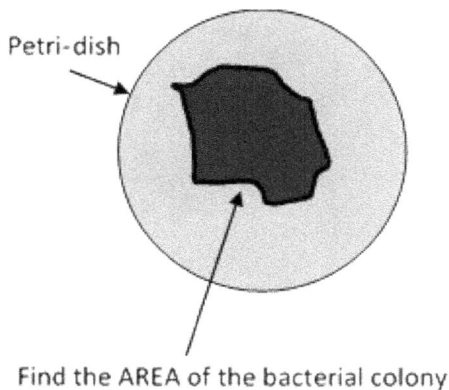

Figure 8.5 Illustration of a bacterial colony in a petri-dish.

Most of the responses from chemistry teachers were on measurement and *units*. Some examples included: discussions on *graduated cylinders* used in the lab with measurements in millilitres (mL); calculating physical quantities such as *Isolated mass of a substance* determined as the difference between the mass of container and substance and the mass of container and *percentage yield* as the quotient of mass of product to mass of initial substance multiplied by 100. Most responses from the biology teachers were on *estimation*. Some examples included: discussions on calculating *population estimates* for a geographical region from a count of sample population collected from a region; pigment chromatography; carbon-14 age determination; using mathematics as they plot graphs to help them understand equations.

After some rich discussion and having them appreciate the application of mathematics in their field, the problem in Figure 8.5 was launched for discussion. The teachers were asked to work in groups to figure out the area of the non-standard geometric shape of the bacterial colony. As a first step, they were given the radius of the circle that represented the petri-dish, which helped them to find the exact area of the petri-dish by using the area of a circle formula.

To help them with their mathematical thinking, graph paper of various grid sizes was also provided at their tables. With some guided instruction, the teachers were asked to place the graph paper with the biggest grid size on top of the picture (see Figure 8.6). Following that, they were guided to perform some computations. This consisted of finding the approximate number of full squares for the non-standard shape and the circle (representing the petri-dish). Following this guided problem-based learning part, the teachers had

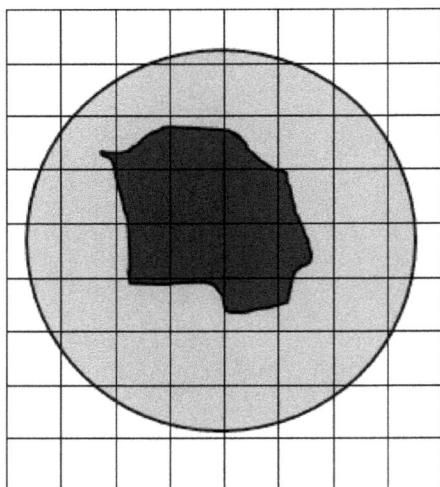

Figure 8.6 Computing area using graph paper with a large grid size.

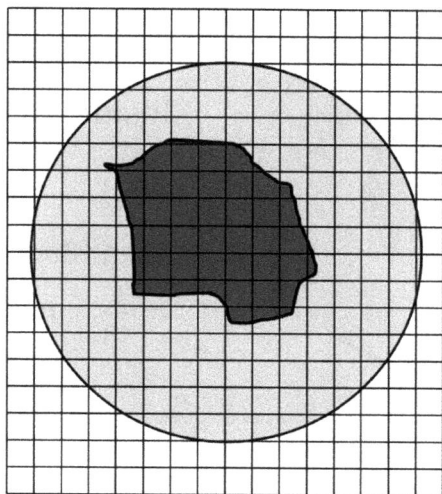

Figure 8.7 Computing area using graph paper with a finer grid size.

the discovery moment when they realized that they can use the respective counts they obtained with the area of the petri-dish that they have and use proportional reasoning to estimate the area of the non-standard shape.

As a continuation, the teachers were asked to repeat the activity using graph paper with a finer grid (see Figure 8.7). When asked what they observed, they immediately noted that the answer they obtained with their proportional reasoning argument this time would be much closer to the exact area of the non-standard shape. Without prompting they had figured out that finer grid papers can yield much better estimates of the area. Needless to say, the teachers were excited with their new mathematical discovery and wanted to go back to the lab to check their answer against the spectrophotometer.

As an added activity in this interdisciplinary problem-based learning experience, we asked the teachers to work in groups to model dropping a pencil onto the sheet with the picture of the petri-dish with the non-standard area. They were asked to keep a count of how many times the pencil fell within the area and how many times the pencil fell outside the non-standard shape but within the circle as shown in Figure 8.8. We note that in the illustration in Figure 8.8, the pencil drops six times within the non-standard shape and seven times outside the non-standard but within the circle. This immediately helps to estimate the area of the bacterial colony as a fraction of the area of the petri-dish as shown.

Even before prompting, the teachers realized that they can continue this process and went on to try dropping the pencil more times and recording the fraction from the number of times the pencil fell within the non-standard

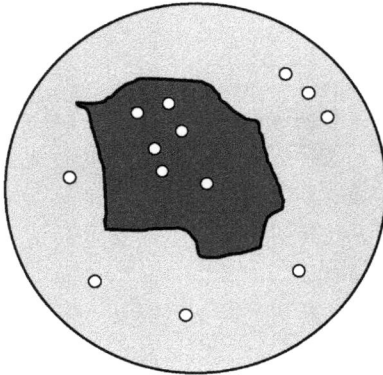

$$\frac{\text{Inside}}{\text{Inside} + \text{Outside}} = \frac{6}{13}$$

Area (Bacterial Colony)
$$= \frac{6}{13} \times \text{Area (petri-dish)}$$

Figure 8.8 Estimating the area of bacterial colony as a fraction of petri-dish area.

shape and the number of times it fell outside but within the circle (see Figure 8.9).

This interdisciplinary problem-based strategy combined effectively with a mathematical modelling framework helped the teachers to develop a mathematical competency of estimating area of non-standard shapes. The biology and chemistry teachers were thankful for that, and they now had more appreciation for applications of mathematics that help them not to simply depend on instruments such as the spectrophotometer. In fact, two teachers mentioned that the instrument in their school stopped functioning and this slowed the functions in the lab. They were very excited that now they can teach their colleagues and other students how to estimate the area efficiently even if they do not have access to the instrument.

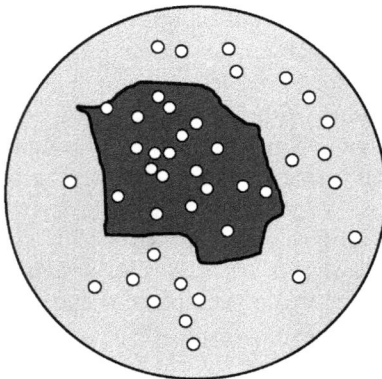

$$\frac{\text{Inside}}{\text{Outside} + \text{Inside}} = \frac{20}{41}$$

Area (Bacterial Colony)
$$= \frac{20}{41} \times \text{Area (petri-dish)}$$

Figure 8.9 Estimating area using geometric probability with more points.

Discussion and conclusion

A summary of the tasks and related educational tools presented in this chapter is given as follows.

Specifically, as shown in Table 8.2, this chapter presented four separate tasks that employed four separate educational frameworks and strategies that help motivate different mathematical competencies. Not only does this help students (and teachers) to understand the content needed to solve the problem but also helps build the necessary mathematical competencies needed to solve the real-world problem in a meaningful and efficient way.

We live in an era where teachers need to understand that 'Students are not just *consumers of education*, but they are *producers of information*'. So our job as a teacher is to become better at *facilitating* student engagement, *anticipating* misconceptions students may have, *monitoring* discussions, *scheduling* student collaborations, *sequencing* critical communications and finally make meaningful *connections*. This is how one can move from a *teacher-centred* classroom to a *student-centred* classroom. These practices will not only help with reasoning, communication and connections but will also help create engaging mathematical tasks, classroom discourse and connections within and beyond mathematics.

We hope that the tasks described in this chapter, the educational strategies and frameworks presented to address them and the related mathematical competencies developed will serve as a resource for practitioners in education to upskill their pedagogical practices and help create a proficient mathematics education workforce.

Table 8.2 A summary of the tasks and related educational tools presented

Task	Find the Height of the Tree
Strategy	*Experiential learning*
Framework	Active learning with notice and wonder approach
Mathematical competency	Ratio and proportion
Task	**What Happens at the Height of the Fountain?**
Strategy	Inquiry-based learning
Framework	5E-Instructional approach
Mathematical competency	Direct and inverse variations
Task	**How many Popcorn Kernels will Fill the Room?**
Strategy	Challenge-based Learning
Framework	FERMI Problem-Solving Approach
Mathematical Competency	Competency: Calculating Volume
TASK	**Find the Area of the Bacterial Colony**
Strategy	Interdisciplinary problem-based learning
Framework	Mathematical modelling approach
Mathematical competency	Estimating area

Acknowledgement

This work was supported in part by a grant from the National Science Foundation DMS 1441024. The author is grateful for this support. This chapter is the result of multiple mathematical education workshops, symposia and seminars conducted by the author in several developing and developed countries, which was presented at the Mico International Mathematics Teaching Summit March 25–27, 2019, Kingston, Jamaica.

References

Ärlebäck, J. B., & Bergsten, C. (2013). On the use of realistic Fermi problems in introducing mathematical modelling in upper secondary mathematics. In *Modeling students' mathematical modeling competencies* (pp. 597–609). Dordrecht, Springer.

Freeman, S., Eddy, S. L., McDonough, M., Smith, M. K., Okoroafor, N., Jordt, H., & Wenderoth, M. P. (2014). Active learning increases student performance in science, engineering, and mathematics. *Proceedings of the National Academy of Sciences, 111*(23), 8410–8415.

Gormally, C., Brickman, P., Hallar, B., & Armstrong, N. (2009). Effects of inquiry-based learning on students' science literacy skills and confidence. *International journal for the scholarship of teaching and learning, 3*(2), n2.

Keeton, M. T., & Tate, P. J. (Eds.). (1978). *Learning by experience–what, why, how* (No.1). Jossey-Bass.

Keselman, A. (2003). Supporting inquiry learning by promoting normative understanding of multivariable causality. *Journal of Research in Science Teaching, 40*(9), 898–921.

Meyers, C., & Jones, T. B. (1993). *Promoting active learning. Strategies for the College classroom, 350* Sansome Street, San Francisco, CA 94104, Jossey-Bass Inc., Publishers.

Pedaste, M., Mäeots, M., Leijen, Ä. & Sarapuu, T. (2012). Improving students' inquiry skills through reflection and self-regulation scaffolds. *Technology, Instruction, Cognition and Learning. (9)* 81–95.

Pedaste, M., Mäeots, M., Siiman, L. A., De Jong, T., Van Riesen, S. A. N., Kamp, E. T., Manoli, C. C., Zacharia, Z. C., & Tsourlidaki., E. (2015). Phases of inquiry-based learning: Definitions and the inquiry cycle. *Educational Research Review, 14*, 47–61.

Peter-Koop, A. (2004). Fermi problems in primary mathematics classrooms: Pupils' interactive modelling processes. In I. Putt, R. Faragher, M. McLean, & Mathematics Education Research Group of Australasia (Eds.), Proceedings of the annual conference of the Mathematics Education Research Group of Australasia (MERGA): Vol. 27. Mathematics education for the third millennium: Towards 2010. Proceedings of the 27th annual conference of the Mathematics Education Research Group of Australasia, Townsville, (pp. 454–461). Sydney, MERGA.

Polya, G. (1957). *How to solve it.* Originally published in 1945 by Princeton University Press.

Rumack, A. M., & Huinker, D. (2019). Capturing mathematical curiosity with notice and wonder. *Mathematics Teaching in the Middle School, 24*(7), 394–399.

Sternberg, R. J. (2008). Interdisciplinary problem-based learning: An alternative to traditional majors and minors. *Liberal Education, 94*(1), 12–17.

Wood, D. F. (2003). Problem based learning. *BMJ: British Medical Journal, 326*(7384), 328–330.

Sentance, S., Dorling, M., McNicol, A., & Crick, T. (2012, November). Grand challenges for the UK: Upskilling teachers to teach computer science within the secondary curriculum. In *Proceedings of the 7th workshop in primary and secondary computing education* (pp. 82–85).

Sin, K. F., Tsang, K. W., Poon, C. Y., & Lai, C. L. (2010). Upskilling all mainstream teach-ers. Teacher education for inclusion: Changing paradigms and innovative approaches. In C. Forlin (Ed.), *Chapter in teacher education for inclusion: Changing paradigms and innovative approaches* (pp. 236–245). New York, Routledge.

Peters-Burton, E. E., Seshaiyer, P., Burton, S. R., Drake-Patrick, J., & Johnson, C. C. (2015). The STEM road map for grades 9–12. In *STEM road map* (pp. 124–162). New York, Routledge.

Seshaiyer, P., & Kappmeyer, K. (2016). Transforming practices in mathematics teaching and learning through effective partnerships. In *Mathematics education* (pp. 105–120). Cham, Springer.

Chapter 9

Teacher subject knowledge for enhancing learners' mathematical thinking

Shandelene Binns-Thompson

Mathematics education in Jamaica

Jamaica seeks to develop an education and training system that produces well-rounded and qualified individuals who will be empowered to learn for life, able to function as creative and productive individuals in all spheres of our society and be competitive in a global context (Vision 2030 Jamaica National Development Plan, 2009).

As Jamaica strives to take its place within a fiercely competitive and highly globalized marketplace, its school graduates must be equipped with the requisite mathematical knowledge, and must, too, possess reasoning, problem solving and critical thinking skills if they are to support national development and if they are to access the kinds of jobs that are emerging, and compete internationally (National Mathematics Policy Guidelines, 2013).

This vision of Jamaica is however hampered by the under-achievement in Mathematics that now challenges the nation's schools. Performance in Mathematics at both the primary and secondary levels of the education system ranges from poor to average over the last decade. Statistics have shown that from 2009 to 2018 the nation's average in Mathematics on the Grade Six Achievement Test (GSAT) ranged from 53% to 63%. Additionally, for the same period (2009–2018) the average pass rates for the Caribbean Secondary Examination Certificate (CSEC) mathematics examinations ranged from 31.7% to 62%.

Cognizant of this under-achievement in mathematics, Jamaica's Ministry of Education, Youth and Culture in its Mathematics and Numeracy Report is adamant that every mathematics classroom embrace the development of analytical, reasoning and critical thinking skills as one of the central goals of mathematics teaching.

The National Mathematics Policy (2013), however, strongly stipulates that whilst it is important to employ effective pedagogical practices it is important that Jamaican classrooms be endowed with mathematics teachers who possess sound subject knowledge to teach at the assigned level. Teachers need wide and deep knowledge if they are to respond well to the needs of the students during the teaching and learning discourse (French, 2005).

As put forward by Ofsted (2000), a teacher's own lack of understanding in the area of science to be taught can lead them to operate in "safe" mode. This involves telling pupils facts, requiring them to copy notes and avoiding activities that would require discussion and the possible revelation of their own lack of understanding. Ofsted (2000) further outlines that in order for teaching practices to be effective teachers must possess the appropriate subject knowledge. It is in light of this, that this project sought to build Jamaican teachers' mathematical subject knowledge.

Introducing the subject knowledge enhancement (SKE) programme

Through independent research, the Plymouth University SKE programme was discovered as an effective tool for building teachers' mathematical subject knowledge and subsequently improving performance in Mathematics. "Deep subject knowledge" or "understanding mathematics in depth" is widely expressed as an important dimension of teachers' mathematical knowledge for teaching (Krauss, Baumert, & Blum, 2008).

As put forward by Ball, Thames, and Phelps (2008, p. 400) "Teaching requires knowledge beyond that being taught to students, and teachers require what they call 'unpacked' mathematical knowledge which they use to teach decompressed mathematical knowledge". These views have been endorsed by the National College for Teaching and Leadership (NCTL – replaced in 2018 by Department for Education and Teaching Regulation Agency) who posits that SKE programmes are aimed at updating a candidate's subject knowledge so that he or she is ready to teach (NCTL, 2015). It is clear that how effective a teacher is in developing students' understanding of mathematics is based on his/her own understanding of the discipline.

In 2015 connections were made with the Centre for Innovation in Mathematics Teaching (CIMT) at the University of Plymouth which had designed and implemented the SKE platform. The heart-rending state of mathematics education in Jamaica and the prospect of SKE positively impacting Mathematics Education were discussed and an agreement was made to introduce the SKE to the Caribbean with a pilot of 24 teachers from four primary schools in Kingston Jamaica.

The University of Plymouth, through CIMT, granted the Jamaican teachers free access to their SKE platform. The Mico University College through The Caribbean Centre of Excellence (then directed by Mrs. Shandelene Binns-Thompson) was in charge of implementing this SKE programme in Jamaica. The Ministry of Education, Youth and Culture gave formal permission for access to be granted to the Nation's schools. Sterling Asset Management provided a financial grant to assist the local research team with executing this project.

The principals of the prospective participating schools were contacted and sensitized about the project. All principals greeted the project with great enthusiasm and were excited to see the effect such a project would have on their teachers and students.

The project was aimed at investigating the impact of enhancing teachers' subject knowledge on critical thinking. In order to examine this impact, the project sought to provide answers to the following research questions:

1 What effect does the SKE training have on teachers' mathematical knowledge and critical thinking?
2 How does the building of teachers' subject knowledge impact learners' achievement in mathematics?
3 Are there barriers for teachers engaging with SKE?

Implementing SKE

The SKE project was officially launched in Jamaica on 7 December 2015. During the launch, the initial four participating schools from Kingston Jamaica were visited and all the participating teachers were fully sensitized about the SKE programme. Additionally, the participating teachers were orientated on how to access and use the SKE platform. Teachers readily realized that the unique nature of the SKE platform will not only help to build their mathematical subject knowledge but also has interactive tools that they thought would be useful as part of their teaching and learning discourses.

This project employed an experimental mixed method research design. A pre-test and post-probe test related to enhancing teachers' subject knowledge for critical thinking was administered. This project was conducted in two phases and included two different sets of participants who were chosen using convenience sampling. The participants of the first phase were from four primary schools in Kingston Jamaica and included 24 teachers and 179 students. This phase ran from December 2015 to May 2016. The second phase of the project involved 34 teachers and 220 students chosen from representative areas in four other parishes in Jamaica. This phase ran from 2017 to 2018. The treatment associated with this project involved enhancing teachers' subject knowledge as a tool for building critical thinking.

The quantitative data for this project were obtained through the use of pre- and post-probe tests on students' analytical and reasoning skills. Qualitative data were obtained through interviews, classroom observations and survey questions.

Intervention

The intervention involved the training of a total of 58 Jamaican Primary school teachers in SKE. This training was geared at building the teachers' subject knowledge beyond the scope of what they were teaching. The SKE training was primarily done using CIMT's SKE platform; however weekly face-to-face sessions were held with participants to offer content remediation and conceptually build the teachers' understanding of concepts that might pose a challenge for them while using the SKE platform.

The SKE platform which is highly user friendly was accessible to participants at their own conveniences during the project. It allowed the Jamaican teachers to build their competencies in

- Computations
- Number theory
- Measurement probability
- Data analysis
- Algebra
- Relations functions and graph
- Angle geometry
- Geometry and trigonometry
- Transformations vectors and matrices.

The areas covered by the SKE training are fundamental core knowledge embedded in the five strands covered by mathematics education in Jamaica. These strands are:

- Numbers
- Measurements
- Geometry
- Statistics and probability
- Algebra.

During this intervention, the teachers were exposed to detailed content information on each area, interactive learning activities, pre- and post-test materials to examine their growth on each unit studied. Upon the successful completion of the programme, the teachers were certified in SKE by the University of Plymouth.

Evaluation

This section will present the findings from the evaluation, providing answers to the research questions and subsequently, evaluating the impact the project had on Primary Mathematics Education in Jamaica. Our first research question is repeated as follows:

What effect does the SKE training have on teachers' mathematical knowledge and critical thinking?

This research question examined the impact of the SKE programme on two different variables: "teachers' mathematical knowledge" and "teachers'

Table 9.1 Descriptive statistics of the pre- and post-probe test of the participating teachers

	N	Minimum	Maximum	Mean	Std. deviation
Pre-test	58	5	83	58.71	16.590
Post-test	58	50	100	80.71	12.150
Valid N (listwise)	58				

critical thinking". To answer these questions teachers' performance on the pre- and post- probe Mathematics tests, their performance on the SKE final assessments and the responses from the teachers' interview sessions were analysed and represented in the following tables.

The data from the pre- and post-probe tests (Table 9.1) revealed that the mean of the pre-probe test was 58.71% and the mean of the post-probe test was 80.71%. A mean difference of 22 between the pre- and post-probe tests was found to be significant. After the completion of the SKE training 90% of the teachers who completed the programme obtained SKE certification with distinctions, 8% received merits and 2% receive pass certification.

It was also disclosed that before the teachers were trained in the SKE programme, they resorted to using trial and error when solving non-routine math-based problems. After the training, the teachers were more systematic in solving non-routine questions.

Additionally, after the intervention, the teachers were also better able to outline the steps taken to solve non-routine questions (Table 9.2). During the interview sessions, the teachers commented on how they were now better able to systematically dissect a non-routine question and place it into algebraic form for solving. Evidence of this was noted in the sample feedbacks of the teachers outlined as follows:

> ...before the training, I usually rely on trial and error to answer questions I was not sure about... After the training, I developed a better understanding of mathematical concepts and found myself breaking questions apart to solve them
> ...I feel like I am better able to reason out a question now that I have completed the SKE training
> ...I find that I am better able to see what questions are asking... also I am better able to apply my math knowledge to solve questions

So the results indicated that the training of teachers in SKE has positively impacted their mathematical subject knowledge. As implied by the results, the teachers' critical thinking skills were also improved after the training in SKE.

Evidence of this was noted on how the teachers approached the probe questions after the training and how well they explained the steps taken to

Table 9.2 Results from the pre- and post-probe testing

Probe questions	Number of respondents with correct response		Percentage of respondents with correct response	
	Pre-test	*Post-test*	*Pre-test*	*Post-test*
1. In a school, $\frac{5}{8}$ of the students are girls. If there are 135 boys, what is the total number of students at the school?	29	42	50	72
2. How many ways can 8 children be arranged in 3 lines?	17	33	29	57
3. The length of a wall is $(14x + 6)$ feet. A couch is placed along the wall that is $(5x + 4)$ feet long. Find the amount of open space along the wall.	42	53	72	91
4. The figure represents the model of Ms. Jones' flower garden. Given that each dimension is measured in metres, what amount of surface does the garden cover?	35	48	60	83

Probe questions	Pre-test	Post-test	Pre-test	Post-test
5. A garden measuring 12 m by 16 m is to have a pedestrian pathway installed all around it, increasing the total area to 285 m². What will be the width of the pathway?	39	53	67	91
6. For the following diagram, find the length of the side labelled *y*.	44	54	76	93

arrive at answers to the given problems. The responses to the questions on the post-probe test also showed that the teachers were applying their problem solving and analytical skills in solving the problem. The teachers' dispositions towards their mathematical subject knowledge were also impacted by this training. This was confirmed by the responses of the teachers during the interview sessions.

One teacher stated that:

> this training has left me feeling smarter, it helped me to remember things in mathematics that I had forgotten and I have learnt so much new math concepts. I feel more confident to teach mathematics.

Our second research question is:

> How does the building of teachers' subject knowledge impact learners' achievement in Mathematics?

Results from the pre-test and post-test for the participating learners were analyzed and reported in the following text. In year one the mean of the pre-test was 22.98% and the mean of the post-test was 38.89% (Table 9.3). In year one, a mean difference of 15.92 between the pre- and post-tests was found to be significant as it was in year two, where the mean of the pre-test was 23.65% and the mean of the post-test was 42.24%.

The results indicated that learners' academic performance in mathematics was positively impacted when their teachers' subject knowledge was enhanced. The teachers' disposition towards teaching changed after they were trained in SKE. Evidence of this change was noted in the interview sessions with the teachers where they commented on how they feel better equipped to deliver mathematical content. "...*I think I am better able to explain the concepts to my students now...*" stated a teacher during the interview.

Crucially though the learners' disposition towards the subject was also impacted by this training. Based on the classroom observation it was noted that students were more receptive to the teachers when the explanations of concepts were clearer compared the earlier lesson before the teachers were trained. I can vividly recall one learner shouting out "*Miss, mi a understand*

Table 9.3 Descriptive statistics of the pre-test and post-test of the participating students

Participating year	N	Mean % (Pre-test)	Std. deviation	Mean % (Post-test)	Std. deviation
Year 1	179	22.98	21.43	38.89	21.96
Year 2	220	23.65	21.03	42.24	22.67

better now, you a explain more these days…". Another learner agreeing with his peer stated *"A true miss, wi start liking maths now… it nuh suh hard"*

Our third research question was:

Are there barriers for teachers engaging with SKE?

The findings from the teacher interviews were used to answer this question.

Several barriers were identified for training Jamaican Primary School teachers of Mathematics in SKE. These barriers were categorized in three specific areas:

Teachers' fear towards Mathematics: Approximately 70% of the participating teachers exhibited great fear towards mathematics. Based on the answers provided to the interview questions, some teachers fear the subject, some were fearful of failing in the SKE programme and some teachers feared that the researchers identified deficiencies in their mathematical knowledge. This was evident in the comments the teachers made while they were being interviewed. These comments included:

> … I have not done math since High school, I just do not feel comfortable doing the subject
> …What if I am unable to do the Math in SKE.
> …What if I fail at SKE?
> … I am afraid that you will notice that I am not good at Math?

Learning Barriers: Some of the participating teachers had limited access to technological devices and the internet. This limited the teachers to completing SKE activities while they were at work, which for the teachers was a challenge since their work-based obligations were great. Another learning barrier that presented itself is that some teachers were not au fait with using technology. That is, these teachers were unable to effectively use the basic functions of the computer, which hampered their ability to interact with the SKE platform.

Time-Constraint Barriers: Participants complained that it was hard to find time to complete the activities and assignments assigned to them on the SKE platform. The teachers stated that the demands of work and home were negatively impacting their ability to effectively pace themselves to complete SKE activities in a timely manner. A sample of the following teachers' feedback indicates some of this evidence.

> …I struggled with finding the time to complete the activities on SKE… It was hard to prepare my students for their GSAT exam and complete the SKE training at the same time
> … My biggest struggle while taking the SKE course was finding the time to complete all the activities…

...As a senior teacher and a mother it was hard to complete the work on SKE and balance everything else...

Conclusions

Results of this intervention confirmed that the training of Jamaican Primary School teachers of Mathematics in SKE has positively impacted Mathematics Education in the participating schools. The participants and their students' academic attainment in mathematics were positively impacted by the programme. The means of both the post-probe test and the student based post-test were significantly greater than those of the pre-tests ($p < 0.05$).

This finding is supported by large-scale studies that reported that there are positive correlations between teachers' subject knowledge and students' achievement (Baumert et al., 2010; Hill, Rowan, & Ball, 2005). This view is supported by Metzler and Woessmann (2010) who posited that teachers' subject knowledge exerts a statistically and quantitatively significant impact on student achievement.

Whilst we uncovered a number of barriers that made participation difficult, namely:

- Teachers' fear of mathematics;
- Limited access outside school to the required technology;
- Time constraints for busy working teachers.

Our evidence though indicates that Mathematics subject knowledge is important and that enhancing this at the primary level does indeed make a significant impact with enhanced progress of the teachers' learners.

Whilst we recognize that enhanced mathematical knowledge does not guarantee that you are a great teacher of mathematics, we do think it a necessary condition for teachers to have to be a great teacher, particularly teachers in the early primary years. It enables you to:

- Better understand of the misconceptions made by your learners;
- Give alternative strategies for learners who have not understood a concept;
- Generalise and extend topics to motivate your learners;
- Enthuse your learners as you know where the topic under study is leading to and hence its importance for learners in making mathematical progress;
- Have confidence and capability to pursue different approaches and solutions that your learners might make;
- Help your learners to become mathematical thinkers.

The countries that have shown enhanced levels of mathematical progress at the secondary level have ensured that their learners develop not just numeracy at the primary level but also the beginnings of mathematical thinking, including early algebra. It is paramount for developing countries to put efforts into enhancing mathematics teaching in the early years of education to ensure that learners have a mathematical foundation which can be built on in years to come and for this to happen, we need out teachers to have enhanced subject knowledge.

Even though this is a long-term strategy for success, it is a vital one for enhanced mathematical progress in developing countries.

References

Ball, D., Thames, M., & Phelps, G. (2008). Content knowledge for teaching: What makes it special? *Journal of Teacher Education, 59*(5), 389–407.

Baumert, J., Kunter, M., Blum, W., Brunner, M., Voss, T., Jordan, A., et al. (2010). Teachers' mathematical knowledge, cognitive activation in the classroom, and student progress. *American Educational Research Journal, 47*(1), 133–180. doi:10.3102/0002831209345157.

French, D. (2005). Subject knowledge and Pedagogical Knowledge. Retrieved from http://www.maths.manchester.ac.uk/~avb/pdf/DougFrenchSubjectKnowledge.pdf

Hill, H. C., Rowan, B., & Ball, D. L. (2005). Effects of teachers' mathematical knowledge for teaching on student achievement. *American Education Research Journal, 42*(2), 371–406.

Krauss, S., Baumert, J., & Blum, W. (2008). Secondary mathematics teachers' pedagogical content knowledge and content knowledge: Validation of the COACTIV constructs. *ZDM, 40*(5), 873–892. doi:10.1007/s11858-008-0141-9.

Metzler, J., & Woessmann, L. (2010). The Impact of Teacher Subject Knowledge on Student Achievement: Evidence from Within-Teacher Within-Student Variation. Discussion Paper No. 4999 June 2010. Retrieved from http://ftp.iza.org/dp4999.pdf

National College for Teaching and Leadership (NCTL). (2015). Subject Knowledge Enhancement: An Introduction. Retrieved from https://www.gov.uk/guidance/subject-knowledge-enhancement-anintroduction

National Mathematics Policy Guidelines, 2013. https://moey.gov.jm/sites/default/files/National%20Mathematics%20Policy%20Guidelines%20%282013%29.pdf

Ofsted. (2000). Secondary Subject Inspection 1999/2000: University of Hull (Mathematics) London: Ofsted.

Vision 2030 Jamaica National Development Plan. (2009). Planning for a Secure & Prosperous Future. *Planning Institute of Jamaica* Retrieved from www.pioj.gov.jm

Chapter 10

Teaching mathematics creativity and innovation through a STEM basis

Albert Benjamin and Joy Baker-Gibson

Promoting STEM

There is little doubt as to the role of innovation in the growth and development of a nation's economy. Countries such as Singapore, Japan, China and South Korea have placed a premium of technological innovation as a corner stone for national growth and development. These nations rank well in the Programme for International Student Assessment (PISA) in Mathematics and Science. Science, Technology, Engineering and Mathematics have at their core problem solving and innovation strategies. Proficiencies in these STEM subjects have been placed at the centre of national development for countries like those previously listed. This paper argues that for countries such as Jamaica and others in the Caribbean to realise the levels of innovation required for 21st-century economies, a national/regional STEM educational ecosystem approach is required similar to that in the United States and India. A model for a national STEM ecosystem in Jamaica is proposed.

Low tide of STEM output

The expanding array of pre-tertiary and higher education offerings within Caribbean nation states has not kept pace with the required output of STEM professionals needed to support modern 21st-century economies. The United States Department of Commerce estimates that over the next decade STEM jobs in the United States will grow by 17%, compared to an estimated 9.8% for other occupations.

The Global Innovation Index (GII) that ranks 125 countries based on assessment of an array of indicators for innovation, ranks Jamaica 81 with only one other country from the region, Trinidad and Tobago, appearing on the index at 96. In relation to innovative outputs from universities, Jamaica ranks better at 51 with Trinidad at 102 out of a total 119 (Global Innovation Index, 2018). Taking a closer look at the case of Jamaica in the performance at the Caribbean Secondary Education Certificate Examinations (the terminal standardised examination taken at high school) we see a tide of low outputs from this pre-tertiary phase of education in the STEM subjects:

- The current sitting pass rate of mathematics is 62%. This represents a 24% increase in the pass rate since 2012 but represents less than 40% of the Grade 11 cohort exiting the secondary education system. This means that at current levels of performance, a significant number of the Jamaican workforce will lack essential mathematical skills and competencies.
- There is an inadequate number of trained mathematics and science teachers available to the system (less than 15% of the 1800 deployed to the secondary system is fully qualified).
- There is inadequate access to quality science education at the secondary level; only about 5000 of the average 40,000 students sitting CSEC subjects annually take science exams.

STEMing the tide

If we in the Caribbean have learnt anything from the United States about how to promote coordinated innovation and creativity in science and technology, it is that wars and rumours of wars represent the tipping point over and above any robust initiatives coming from systems of education. World War II in 1945 and the Russian launch of Sputnik in 1957 heralded unprecedented developments in the understanding of atomic energy and space exploration for the United States. The birth of NASA itself and the entire western, modern STEM movement in 1958 was as a direct response to the Russian launch of Sputnik in the previous year. Congress passed the "Space Act" in 1958 establishing the National Aeronautics and Space Administration (NASA). The mission of NASA was to "expand and improve" (Dick, (2008) the United States' presence in space through science and engineering. It could be said that subsequently, NASA has been the epicentre of the development of the "STEM career" initiatives in this hemisphere.

Concomitantly with the establishment of NASA was the realisation of the need for a system of education that would promote an increase in the number persons entering the STEM careers (scientists, engineers and mathematicians). A second American organisation called the National Science Foundation (NSF) was instrumental in establishing the "STEM education" agenda. The NSF initiative emphasised a more broad-based effort to promote critical thinking, creativity and problem-solving skills among the general workforce and not just the promotion of an increase in the number of persons pursuing "STEM careers". This initiative was originally referred to as SMET (Science, Mathematics, Engineering and Technology) (Sanders, 2009). The SMET education concept, later revised to STEM in 2001, was launched in the 1990s, over 40 years after the founding of the NSF in 1950.

Unfortunately, the term STEM is often used by itself and this can create some conflation between discrete career initiatives and education initiatives. Contrasted with the NASA mission of increasing numbers of persons entering STEM careers, STEM education attempts a bold and ambitious trans-disciplinary ideal where the core creative operations in Science, Technology, Engineering and Mathematics are extracted and taught to promote critical thinking, creativity, innovation and problem solving to all learners. However discrete, what both NASA and NSF STEM agendas have in common is that they are active components of a national legislative STEM ecosystem agenda.

Another example of effecting legislative frameworks, a prequel to a productive national STEM ecosystem agenda, was also seen in India. It could be argued that among the Commonwealth countries, India perhaps shows a model where wars and rumours of wars had more of a tangential impact but an impact none the less in their advancements in science and technology. In the same year that keeping up with the Russians spurred congress into action in the United States, the Indian Parliament tabled its National scientific policy, which among other things indicated that (India National Science Policy, 1958, pp. 1–2):

> The key to national prosperity, apart from the spirit of the people, lies, in the modern age, in the effective combination of three factors, technology, raw materials and capital, of which the first is perhaps the most important, since the creation and adoption of new scientific techniques can, in fact, make up for a deficiency in natural resources, and reduce the demands on capital. But technology can only grow out of the study of science and its applications…
>
> …The Government of India have accordingly decided that the aims of their scientific policy will be to foster, promote, and sustain, by all appropriate means, the cultivation of science, and scientific research in all its aspects pure, applied, and educational; to ensure an adequate supply, within the country, of research scientists of the highest quality, and to recognize their work as an important component of the strength of the nation; to encourage, and initiate, with all possible speed, programmes for the training of scientific and technical personnel, to ensure that the creative talent of men and women is encouraged and finds full scope in scientific activity; to encourage individual initiative for the acquisition and dissemination of knowledge, and for the discovery of new knowledge, in an atmosphere of academic freedom; and in general, to secure for the people of the country all the benefits that can accrue from the acquisition and application of scientific knowledge.

Even with the established pedigree in STEM areas India still required a national policy framework to ensure sustained progress. Out of this policy framework emerged the Homi Bhaba Center for Science Education (HBCSE) and The Indian Institute of Technology (IIT), Delhi. These two institutions have been at the centre of implementation of the national scientific policy. HBCSE is responsible for research and development in STEM education.

IIT on the other hand is responsible for ensuring a consistent supply of persons for STEM careers. India actually has the highest share (36.2%) of international students enrolled in STEM fields in United States universities, mostly in engineering (World Education News and Reviews, 2018).

What will be the event(s) for individual Caribbean nation states or CARICOM as a whole to create the urgency required to launch our own legislative changes necessary to promote a comprehensive STEM ecosystem agenda? Will it be the rapidly closing borders of the United States or the ever-increasing havoc of climate change?

Perhaps some combination of these along with the current demand to operate as modern digital economies will be our tipping point? It is important to note that whatever the tipping point, a legislative agenda to support the development of institutions charged with implementation of a STEM ecosystem represents the necessary precursor to significant progress.

Our progress as a region with respect to a collective STEM agenda is best summed up by this excerpt from the 39th CARICOM Heads of Government Meeting in Montego Bay Jamaica in July 2018:

> As a region, we must also hasten our steps toward the transition to a Green and Blue economy; understanding that our future development is intrinsically linked to how successfully we exploit these formerly untapped resources. We also need renewed and continuous engagement with our youth populations. Colleagues, in navigating the new normal, and engaging our youth there is one sure vehicle to take us forward, which we are woefully behind in riding. That is STEM (Science, Technology, Engineering and Mathematics) and perhaps at its simplest level—ICT—the Internet and information highway.
>
> We, as Heads of Government, therefore have to be the ones to place STEM and ICT at the heart of what we want to accomplish as a region. Indeed, it appears that every single day, there is some new global development in the technology field. In October 2017, the world gasped when Saudi Arabia became the first country to grant citizenship to a robot; yet we continue to grapple with the rights of our own human CARICOM citizens. (Mitchell, 2018)

Mitchell (2018) bemoans that we are "woefully behind" in accomplishing a cohesive regional STEM career or STEM Education agenda. This outcome is clearly at the heart of why the region is viewed as peripheral consumers rather than significant creators or innovators of science and technology. Without any competitive showing as innovators and creators of technology our opportunities now lie in ensuring that our people can access the emerging 21st-century jobs. This of course implies a need to ensure that we are equipped with the relevant competencies and skills according to the World Economic Forum (2018).

To respond to these demands, changes in societal norms will need to be accommodated. These changes of necessity will depend in large measure on leveraging of the pre-tertiary and higher education priorities.

Such priorities will have to be set in relation to the need for rapid workforce upskilling and expansion in the STEM fields. Before this however, a legislative framework establishing the trophic levels of a **STEM ecosystem** with clear institutional parameters will be ground zero for any successful STEM effort, as was discussed before in the cases of the United States and India.

As only Jamaica and Trinidad and Tobago from the region appear in the Global Innovation Index I will confine the rest of this paper to discussing the case higher up on the list; that of Jamaica with the anticipation that other countries can profit from lessons learnt in attempting to implement a STEM ecosystem in that country.

The Jamaica national STEM ecosystem

Jamaica has a population of about 2.7 million. The country's poverty rate increased by 8% after the 2008 global economic crisis. The *Statistics Institute of Jamaica* estimates that the unemployment rate in Jamaica is now over 13.7%; the average unemployment rate for women is almost double that for men: 18.6% versus 9.6%. In its 2011 *Jamaica Country Economic Memorandum: Unlocking Growth*, the World Bank noted that one of the main causes of Jamaica's low economic productivity is "Deficiencies in human capital and entrepreneurship due *inter alia* to quality deficiencies in education and training and high migration rates for tertiary graduates and skilled workers."

A national priority of the Jamaica Economic Growth Council (the body tasked by the government to facilitate growth in Jamaica) is the development of human capital. One key indicator is the production of at least **6000 engineers to drive economic growth** within the next 5 years.

This ambitious goal is one of the key deliverables of an effective national STEM ecosystem. In addition to engineers, an effective STEM ecosystem must be responsive to general future employment trends.

Table 10.1 depicts a prediction of the relevant future skills and jobs available for persons in the Caribbean and Latin America:

While there is some access to the integrated STEM education model in Jamaica a cohesive STEM ecosystem is not in place to ensure systematic permeation. The 1990 National Science Policy positioned the government of Jamaica as an enabler for development through science and technology.

This policy while far reaching and extensive stopped short of legislating into existence an institution with responsibility to drive the national STEM education research in STEM teaching and learning. In addition, the policy failed to support specific curricular framework for the promotion of a

Table 10.1 Current trend in skill and jobs for Caribbean and Latin America; World Economic Forum, (2018)

Emerging skills required	Emerging jobs
Analytical thinking and Innovation	Software and Application Developers and
Creativity	Analysts, Managing Directors and CEOs
Originality and initiative	Data analytics and Statistics
Active learning and learning strategies	General and Operations Managers
Technology design and programming	Sales and Marketing Professionals
Reasoning, problem solving and ideation	Sales Representatives
Critical thinking and analysis	Wholesale Manufacturing
Leadership and social influence	Technical and Scientific Products
Complex problem solving	Financial and Investment Advisers
Emotional intelligence	Financial Analysts
Resilience, stress tolerance and flexibility	HR Specialists
	Assembly and Factory Workers

national culture of STEM-based learning and problem solving. Figure 10.1 depicts a proposed model for such a national STEM ecosystem. The institutions not shaded are present; those with lighter shading are in need of further technical, legislative and financial support and the areas with the darker shading do not presently exist.

Summary

Technological innovation has demonstrated significant promise for national growth and development in economies such as South Korea, China and Singapore. These countries rank consistently well in the PISA Mathematics

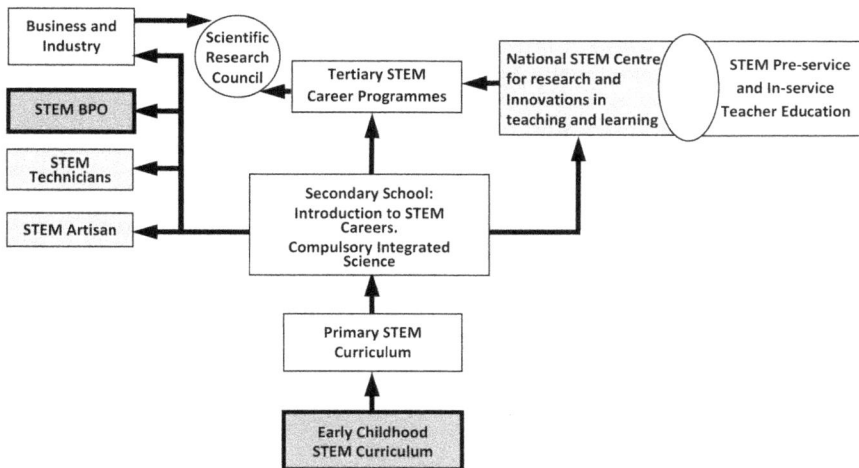

Figure 10.1 Proposed national STEM ecosystem.

and Science assessment. While Jamaica and other countries in the Caribbean do not lack mature educations systems, there is an absence of country level or regional coordination designed to optimize outcomes in the STEM subject areas.

The United States and India have managed to maintain significant output of STEM professionals due in significant measure to their use of a national educational STEM ecosystems approach. Jamaica and countries in the region should consider using this model as a basis for increasing the amount of STEM professionals and in support of overall innovation for economic development.

References

Dick, S. (2008). The Birth of NASA. Available at: http://www.nasa.gov/exploration/why-weexplore/Why_We_29.html. Retrieved April 7, 2019.

Global Innovation Index. (2018). Cornell University, INSEAD, and the World Intellectual Property Organization, 2018. Available at: https://www.globalinnovationindex.org/Home. Retrieved April 15, 2019.

Mitchell, K. (2018). CARICOM Today. Available at: https://today.caricom.org/2018/07/06/place-stem-ict-at-heart-of-what-we-want-to-accomplish-pm-mitchell/. Retrieved May 9, 2019.

National Labour Market Survey. (2017). A Guide to Employment Opportunities in Jamaica. Published by the Ministry of Labour and Social Security, Kingston Jamaica, West Indies. Available at: http://www.lmis.gov.jm/common/ViewDocument/518e8201-c536-4daf-b54e-213e37586fb3. Retrieved April 15, 2019.

Sanders, M. (2009). STEM, STEM education, STEMmania. *The Technology Teacher*, 68(4), 20–26.

World Economic Forum. (2018). Future of Jobs 2019 Latin America and the Caribbean. Available at: http://reports.weforum.org/future-of-jobs-2018/latin-america-and-the-caribbean/. Retrieved April 5, 2019.

World Education News and Reviews. (2018). Education in India. Available at: https://wenr.wes.org/2018/09/education-in-india. Retrieved April 5, 2019.

Chapter 11

Lesson study

Generating, accumulating and sharing professional knowledge

Derek Robinson

Background

In the McKinsey report, *How the World's Best-Performing School Systems Come Out on Top*, Barber and Mourshed (2007) examined 25 school systems worldwide, including the top ten performing systems, to determine the common factors necessary to improve student performance. The authors concluded that successful schools need to do three things: get the right people to be teachers; get the best out of these teachers and step in when pupils start to lag behind.

They went on to say, "The quality of a school system rests on the quality of its teachers", and added, "The top-performing school systems recognise that the only way to improve outcomes is to improve instruction". As Hopkins (2007) had earlier stressed, "it places the focus of reform directly on enhancing teaching quality and classroom practice rather than on structural change". Guskey (2000) effectively summarised this message by saying, "Never before in the history of education has greater importance been attached to the professional development of educators".

Barber and Mourshed go on to say that the challenge faced by the government, universities, schools and teachers is to firstly define what "great instruction" looks like and secondly, "to give teachers the capacity and knowledge to deliver great instruction reliably everyday across thousands of schools that vary enormously".

One of the recommendations in the McKinsey report was that schools should look at Japanese lesson study as a means of enabling teachers to work together and learn from each other.

At the time, little was really known about Japanese lesson study in the United Kingdom, and, although this situation has changed over the past 15 years, there is still a concern amongst Japanese educators that "many aspects of lesson study that are well understood by Japanese teachers have not transferred readily [to other countries" (Fujii, 2013).

Over the past 15 years my department and I have been working with Centre for Innovation in Mathematics Teaching (CIMT) in developing creative approaches to on-going forms of collaborative professional development. Initially we concentrated on the Hungarian approach to professional development through University Practice schools with CIMT acting as our advisers. However, the current trend in the United Kingdom is for schools to take control of their own professional development and this eventually led us to look more closely at Japanese lesson study as a possible alternative. Japanese lesson study first began to attract widespread interest in the West following the publication of The Teaching Gap, written by Stigler and Hiebert (1999), which offered compelling evidence of the effectiveness of lesson study as a tool for professional development.

Stigler (2000) states that, "The key to long-term improvement is to figure out how to generate, accumulate, and share professional knowledge". Isoda, Stephens, Ohara, and Mijakawa (2007) believe that lesson study does exactly that.

First visit to Japan

In order to understand more about lesson study, Professor David Burghes of CIMT, arranged for a group of teachers to visit Japan in the summer of 2006 where we were able to observe several research lessons and listen to the post-lesson discussions. Although we were given translations of the lesson plans, we were unable to follow much of these discussions. Nevertheless, the trip was enough to inspire me to introduce lesson study into the mathematics department at my school on my return.

During the next few years lesson study became my department's main focus for the professional development of the whole team. This brought the department much closer together, made watching other people's lessons the norm and enabled us to jointly plan and review many lessons. During this period, our teaching was largely based on the Hungarian approach to teaching that underpinned CIMT's MEP KS3 course. This approach to teaching is based around whole-class interactive teaching, with its emphasis on continually monitoring each pupil and encouraging contributions from as many pupils as possible, including working at the board in front of the class. Most lessons have clear lesson objectives, are fast-paced and contain many episodes often lasting less than 10 minutes. This meant that in our scheme of work "problem solving" was generally used to practise a previously taught skill and to develop general problem-solving skills. It was seldom, if ever, used to introduce and develop new concepts (Van de Walle, Karp, & Bay-Williams, 2010).

Having worked through several cycles of lesson study, we gradually began to realise that although we were "accumulating and sharing" professional knowledge, perhaps we were not generating anything particularly new. In effect, we were using lesson study to share and improve our Hungarian style of teaching. We were not alone in this, as, although many teachers have taken the lesson study process to heart, not everyone has understood the significance of the related Japanese structured problem-solving approach (Doig, Groves, & Fujii, 2011). As a department, we felt we had to continue with our search for "great instruction" and how we might look implement it in our classroom.

Teaching through problem solving

Current mathematics education reforms both here in the United Kingdom and abroad suggests that, "problem solving and investigative approaches are central to learning for all pupils" (Ofsted, 2011). It is certainly true in many of the countries the UK government wish to emulate.

For example, in Singapore, mathematical problem solving is central to mathematics learning (Singapore Ministry of Education (MOE), 2007). It involves the acquisition and application of mathematics concepts and skills in a wide range of situations, including non-routine, open-ended and real-world problems. The development of mathematical problem-solving ability is dependent on five interrelated components, namely, Concepts, Skills, Processes, Attitudes and Metacognition (Singapore MOE, 2007).

And in Japan, the course of study (COS) outlines the importance of developing new mathematical activities that inspire the students to "willingly engage in mathematics with purpose" by "trying to find new properties or to create new ways of thinking or to solve concrete problems" (Japanese Ministry of Education, 2008). Isoda and Katagiri (2012) summarise this by reporting that the basic principle of the problem-solving approach is to nurture children's learning of mathematics by/for themselves.

This is not a new idea as back in 1980 the National Council of Teachers of Mathematics (NCTM) suggested that "problem solving must be the focus of school mathematics" (NCTM, 1980, p. 1). It concluded with the publication of Everybody Counts (National Research Council, 1989) and the Curriculum and Evaluation Standards for School Mathematics (NCTM, 1989), both of which emphasise problem solving.

This was certainly the case in the United Kingdom, where paragraph 243 of the *Cockcroft Report* (1982) recommends that Mathematics teaching at all levels should include opportunities for:

- exposition by the teacher;
- discussion between teacher and pupils and between pupils themselves;
- appropriate practical work;
- consolidation and practice of fundamental skills and routines;
- problem solving, including the application of mathematics to everyday situations;
- investigational work.

As with all new initiatives, what might have been clear to the architects of the report, is not necessarily as clear by the time it reaches the classroom as this is influenced by various other factors. Firstly, most teachers have their own beliefs as to the 'correct way of teaching maths'; these beliefs being largely shaped "through informal participation over long periods of time". In other words, most prospective teachers are heavily influenced by their own experience at school and consequently teach in a way how they themselves were taught (Bartley, 2007). Secondly, and increasingly more so, teachers are influenced by the examination boards and performance targets (Cambridge Assessment, 2012). There are, however, definite echoes of Cockcroft in the Ofsted (2011) guidelines to schools, which re-emphasises that "problem solving, discussion and investigation are seen as integral to learning mathematics".

It therefore comes as no surprise that the Statutory Programmes of Study for Mathematics (DfE, 2013) have placed a greater emphasis on problem solving by stating that our students should be "able to solve problems by applying their mathematics to a variety of routine and non-routine problems with increasing sophistication, including breaking down problems into a series of simpler steps and persevering in seeking solutions". Although this is not a new idea, as seen earlier, what followed certainly is for many of our teachers as they are told to ensure that their students "develop their mathematical knowledge, in part through solving problems and evaluating the outcomes, including multi-step problems".

If there is general agreement that problem solving should be at the heart of mathematics teaching (Schoenfeld, 1992) then why has progress been so slow?

Pehkonen (2008) explains that the apparent lack of progress is caused by a lack of clarity in both the meaning and goal of problem solving as there is an international recognition that problem solving is a "fuzzy concept" (Pehkonen, 2008). For example, Schroeder and Lester (1989) identified three types of approach to problem solving as shown in the following text:

> **Type 1 – Teaching for problem solving.** This generally means the teacher teaches a specific topic and then sets "problems" to practise the new skill. This was the American approach highlighted by Stigler and Hiebert in their report. It still is the approach adopted by most textbooks in the United Kingdom.
> **Type 2 – Teaching about problem solving.** This approach involves teaching students how to problem solve by identifying strategies that might be useful in approaching a task. For example, this could include: draw a picture; look for a pattern; make an organised list. Identifying strategies provides a useful means for students to talk about their methods.
> **Type 3 – Teaching through problem solving.** This approach means that students learn mathematics through problem solving. It is almost the opposite of teaching for problem solving with the problem presented at the beginning of the lesson and the skill emerging through working on

the problem during the lesson. This is the Japanese approach highlighted by Stigler and Hiebert (1999).

Over the past 30 years, various initiatives in the United Kingdom have put more emphasis on types 1 and 2 rather than type 3 approaches to problem solving. Consequently, as Tall (2008) points out children in the West spend most of the lesson doing a "large range of examples or problems" whereas in Japan the whole lesson is devoted to a single problem. He stresses that the "Japanese classroom focuses much more on the flexibility of ideas" than on perfecting a particular skill. In other words, "mathematics is taught through problem solving" (Van de Walle et al., 2010).

The problems now faced in the United Kingdom mirror those faced by Finland in the late 1980s and early 1990s when the "National Board of Education made systematic efforts to promote problem solving in school mathematics" (Pehkonen, 2008). Initially in Finland, the lack of "appropriate tasks" was blamed for slow uptake of this approach. This was, however, rectified by a new wave of problem-based textbooks from around 1996 that were focused on teaching through problem solving. Despite the general recognition that these books made "mathematics teaching more interesting", less than 10% of Finnish teachers chose to use these new materials as they "demanded much preparatory work from the teachers' side" (Pehkonen, 2008).

This supports Stigler and Hiebert's (1999) view that teachers tend to follow cultural scripts and their effectiveness depends on the scripts they use. They went on to say, "The fact that teaching is a cultural activity explains why teaching has been so resistant to change".

Stigler and Hiebert (1999, p. 133) suggest that the problem cannot be solved by simply recruiting better teachers as the problem is much deeper. The thrust of their argument is that; "...teachers follow scripts as members of their culture, and their effectiveness depends on the scripts they use" (Stigler & Hiebert, 1999, p. 134).

It is with this in mind that teachers in the West have turned their attention to the Japanese approach to teaching through problem solving as well as their approach to professional development. This, in turn, led me to look more closely at the "Japanese script" and for me to revisit Japan in 2012 with the specific intention of learning more about the Japanese script and "teaching through problem solving". But to really understand the Japanese script, it is necessary to first understand the structure of the Japanese education system.

Japanese educational system

Schools: Japanese schools are divided into three categories: elementary (Grades 1–6, ages 6–12); lower secondary (Grades 7–9, ages 13–15) and upper secondary (Grades 10–12, ages 16–18). Virtually all students go on to upper secondary school, some of which are focused on agriculture, commerce or industrial technology classes. Throughout elementary school, classes are mixed ability.

Teachers: To teach in Japan, one has to hold a teaching license issued by the Prefectural Education Board. This is obtained by completing a teacher education programme at an approved college. Teachers at the elementary schools teach all subjects whilst teachers in secondary schools are content-area specialists. There is no shortage of teachers in Japan but, in most elementary schools, there will only be one or two mathematics specialists who play a major role in helping others improve their mathematics through lesson study.

Sugiyama (2008) distinguished three levels of teaching that Takahashi (2011) summarised as follows:

> Level 1: The teacher can tell students important basic ideas of mathematics such as facts, concepts and procedures.
> Level 2: The teacher can explain the meanings of and reasons behind the important basic ideas of mathematics in order for students to understand them.
> Level 3: The teacher can provide students opportunities to understand these basic ideas and support their learning so that the students become independent learners. (pp. 201–202)

As Fujii (2014) explains, Level 1 teachers provide the "what and how", while Level 2 teachers provide "what, how and why", which corresponds to instrumental and relational understanding in Skemp's (1976) terminology. Level 3 teachers, however, provide students with the opportunity to explore, by themselves, mathematical ideas resulting from their own thinking and understanding.

A Level 3 lesson is student-centred because students explore new concepts, relations, rules, etc., not through spoon-feeding or deliberate coaching by the teacher, but mostly due to their own efforts. Consequently, in a Level 3 teacher's lesson, the bulk of the concepts or ideas originate from the students, and it is their voice that is mostly heard during the lesson. It is here that students, rather than the teacher, become the centre of the teaching-learning process.

Sugiyama (2008) states that for a Level 3 lesson, "teachers must be able to choose a good task, identify the pre-requisite knowledge, and, most importantly, must be able to nurture students to apply their knowledge to a new situation". These are skills that are developed and honed through the school's own Lesson Study programme which underpins professional development in Japan. Stigler and Hiebert (1999) point out that "cultural scripts" are learned, not by intentional study, but "through informal participation over long periods of time".

The Japanese script

Becker and Epstein (2005) explain that although there had been collaboration between US and Japanese teachers and professors from the late 1960s, worldwide interest in Japanese teaching methods probably began in the 1986, ignited by the US–Japan Seminar on Mathematical Problem Solving held, appropriately, at the East-West Centre in Honolulu. This seminar focused on exploring the different philosophies and practices of mathematics teaching and research in both countries and acted as a catalyst to further on-going collaboration between educators from both countries.

Shimada (1997) further explains that in the early 1970s Japanese researchers were involved in a project that focused on "substantiating the effectiveness of open-ended problems ... as a method to evaluate higher-order thinking skills". Their findings on what he describes as "The Open-Ended Approach" to teaching were published in a Japanese book as a proposal for improving mathematics teaching. Collaboration eventually led to a translation of the original Japanese book that became the first Japanese research report translated into English, entitled *The Open-Ended Approach: A New Proposal for the Improvement of Teaching.* (Becker and Shimada, 1997)

Perhaps surprisingly the APEC guide to teaching through problem solving begins with a quote from Polya's famous book, *How to Solve It* (1945), "The purpose of mathematics class is to help students, but not too much and not too little". This means that good teachers should leave the student a reasonable share of the work and not give knowledge and procedures by spoon-feeding. In other words, the Japanese approach is based on "teaching through problem solving".

As Kishimoto and Tsubota (2007) explain, most mathematics lessons in Japan encourage students to take an active role in constructing their own mathematics by communicating with one another; students are encouraged to develop a belief in their own ability to learn and to think. These lessons often feature the "Open-ended Approach". Isoda (2007) explains that, the Open Approach is considered to have three aspects, "Processes are open (various ways of solving the problem)"; "Ends are open (various answers against an open-ended problem)"; "Problems are open (changing and developing problems from a given problem)". Each of these three approaches is underpinned by the belief that students solve problems not only to apply mathematics that they have already been taught but also to learn new mathematics.

What is not immediately apparent from this is the importance of the choice of problem. Becker and Epstein (2005) say that the first criterion is that all students will have "some degree of success with it". They go to say that teachers then "write down all the responses" they expect will come up in the lesson. They describe this approach as "absolutely new to all teachers we have worked with". Lewis (2003) says that in this way teachers "hone their capacity to see the lessons from the students' point of view". She adds that Japanese teachers stress that "developing the eyes to see children" is the most important benefit of lesson study. Fernandez and Yoshida (2004, p. 185) stress that the choice of problem must encourage flexible approaches, and this should be made clear by the way the question is framed. They add that in Japan teachers often emphasise this by asking their students to "think about as many different solutions as possible". Tall (2008) stresses that the "Japanese classroom focuses much more on the flexibility of ideas" than on perfecting a particular skill.

Tall (2008) also explains that new concepts are linked to previous learning by designing a problem that naturally extends the students' learning towards the desired outcome. Students solve the problem using their own "natural ways of thinking" and discuss their solutions to see which methods are better than others (Becker and Epstein, 2005; Tall, 2008). Kishimoto and Tsubota (2007, p. 124) affirm this "has much in common with what is known as social constructivism in the USA and Europe".

As mentioned earlier, Stigler and Hiebert (1999) point out that the Japanese approach to teaching mathematics is very different from the lessons observed in both the United States and Germany. The authors noted that in Japan "the teacher presented a problem without first demonstrating how to solve the problem". In contrast they realised that in the USA, problem-solving lessons generally followed extensive teaching on the skill needed to solve the problem. Their book then stimulated educators' from around the world to rethink their approach to using "problem solving" as a learning tool within mathematics lessons.

One of the main differences between Japan and the West is that the teachers in Japan not only share a common understanding of this approach but that they also use specific pedagogical terms to describe and discuss their roles in the process. In contrast, Hopkins (2007) says that, "we simply do not have robust and sophisticated language in this country for talking about teaching".

Shimizu (2007) states that many Japanese mathematics lessons follow a similar pattern with students working individually or in groups on one problem for the whole lesson. The purpose of these lessons is to establish a link between what is known and what needs to be known. In other words, Japanese teachers use problem solving to generate new knowledge rather than consolidate what has already been taught (APEC, 2003). This use of problem solving to generate new knowledge contrasts with the more traditional use of problem solving still common in most textbooks where students are taught a skill which is then used to solve problems (Van de Walle, Karp, & Bay-Williams, 2010). For example, students learn the algorithm for adding fractions, and once mastered, solve word problems that involve adding fractions.

Takahashi (2016) explains that "Japanese problem-solving lessons usually do not end even after each student finds a solution(s) to the problem". The teacher then orchestrates a discussion that highlights the similarities and differences between the students' approaches. Takahashi adds, "Japanese teachers believe that the heart of the lesson begins after students have come up with a solution(s)".

Shimizu (2007) explains that problem-solving lessons generally consist of four phases as shown in the following text and that Japanese teachers play different roles in each phase (Natsusaka, 2007; Shimizu, 2007):

1 Presenting the problem of the day (5–10 min)
2 Problem solving by the students (10–20 min)
3 Comparing and discussing (10–20 min)
4 Summing up by the teacher (5 min)

In Phase 1, Isoda (2007) explains that the teacher presents the problem in a way that can be easily understand and enable students to know what is expected of them. In planning for this phase of the lesson much thought will have been given to the wording of key questions that will provoke the students' thinking (*Hatsumon*). Fugii (2016) adds that this "specifically excludes any exposition by the teacher about how to solve the task".

In Phase 2, Shimizu (2007) explains that the students think about the problem on their own and try to find solutions of their own. Meanwhile the teacher walks purposefully around the room looking at the students' work, making notes and deciding in which order he will ask the students to present their ideas in phase three (*kikan-shido*). The teacher in general does not guide the students apart from giving hints to students who cannot make progress on their own. In this way the maths being created belongs to the students.

In Phase 3, the teacher asks three to five students who used different methods (or obtained different answers depending on the type of question being used) to explain their approaches to the rest of the class. The teacher remains neutral to the ideas. The students listen to the explanations and try to reach a common understanding of the solutions by discussing the strong and weak points of each approach proposed and identifying what they have in common. This process of "polishing" students' ideas through a whole class discussion (*neriage*) is regarded as crucial to the effectiveness of the whole lesson.

In Phase 4, the teacher summarises the group findings and emphasises the important points addressed in the lesson (*matome*). The teacher will challenge the students with similar or developmental problems for homework.

The students often write down what they have learned in their journals.

What is clear is that all Japanese mathematics teachers have a clear understanding of what they might describe ascribe as "great instruction"

On my return to England, we began to look more closely at the Japanese approach to teaching mathematics through problem solving and, in particular, the rationale behind the choice of tasks. This then created a problem as, although there is a growing source of literature on the lesson study cycle, little attention has been given to the choice and design of the task. Doig et al. (2011) believe that more research is needed into "carefully crafted" tasks that are suitable both for the professional development of the teacher and for use in the classroom.

This indirectly led me to visit Japan a third time as part of the International Math-teacher Professionalization Using Lesson Study (IMPULS) in 2016 (IMPULS, 2016). It was during this trip that I was able to further develop my own understanding of lesson study and the Japanese approach to teaching through problem solving.

A lesson from Japan

Although I had visited Japan twice previously, this was still an eye-opening and breath-taking experience.

Through my previous visits and 10 years of being involved in lesson study in the United Kingdom I had a good understanding of how lesson study functions. I was also aware of how difficult it is for teachers in the United Kingdom to implement some of the big ideas involved, especially the concept of learning through problem solving rather than reinforcing learning with problem solving. Consequently, I was delighted to gain a place on the IMPULS trip and to have the opportunity to learn from two of the world's experts on lesson study, Professors Fujii and Takahashi as well as other educators from around the world.

In all, 34 participants gathered from Australia, Malaysia, Netherlands, Portugal, Singapore, Switzerland, the United Kingdom and the United States to join Lesson Study Immersion Program 2016 from June 20 to June 27. During the program, we were able to visit Tokyo Gakugei University International Secondary School, 2 public elementary schools in Tokyo, 1 public elementary school in Yamanashi and Yamanashi University Model Elementary School, to observe in total seven research lessons and the post-lesson discussions with simultaneous English translation.

The advantages of having unobtrusive real-time translations were invaluable.

As well as observing lessons we were given detailed instruction on lesson study and the Japanese Approach to teaching through problem solving. We were also able to share our thoughts on the lessons and ask relevant questions at our basecamp in Tokyo Gakugei University.

This gave me the opportunity to ask about the use of textbooks in schools, how students use their journals and more specific questions about the lesson planning process.

Prior to our arrival in Japan, we were asked to read various papers written by two of our hosts, Professors Fujii and Takahashi, that were not yet available for general publication. To my delight, *Recent Trends in Japanese Textbooks for Elementary Grades: Supporting Teachers to Teach Mathematics through Problem Solving* (Takahashi, 2016) not only covered my question about the use of textbooks in Japan but also illustrated how the textbooks are used to "support students in developing their note-taking skills throughout the grades".

It was interesting to read how Japanese teachers "use the textbook to teach mathematics" (Takahashi, 2016) and it is through the use of these textbooks that teachers gain the necessary skills to teach through problem solving.

The second paper, *Designing and adapting tasks in lesson planning: a critical process of Lesson Study* (Fujii, 2016) highlighted aspects of lesson study that appeared not to be fully understood outside of Japan and clarified the mysteries behind lesson planning that had puzzled me for some time.

Japanese textbooks

The comments made by Professors Takahashi and Fugii about how textbooks are used to plan lessons, were very enlightening. It was encouraging to know that even in Japan new teachers struggle to develop the skills needed to teach through problem solving without strong collegial support from more experienced teachers.

In particular, it was evident that the textbooks act as a major planning tool for teachers as they contain "resources to assist teachers in their instruction via problem solving and to help students to learn through problem solving" (Takahashi, 2016). Although I had looked at the textbook series, "Mathematics International" (Tokyo Shoseki) over several years, I had not fully understood how the teacher uses these textbooks in planning. I had not really grasped that most chapters begin with a "preparation" page on the left, which either covers everyday mathematics or recalls, "mathematics already studied" (Takahashi, 2016).

This is then followed by the opening problem on the right-hand page. This page acts as a stimulus for the teacher's planning as it includes "a way into the question" as well as possible hints that the teacher might choose to use in his or her planning. Although this sounds very much as if the teacher "teaches the textbook" in fact the teacher uses the "textbook to teach mathematics" (Takahashi, 2016). The teacher studies the textbook including the given hints and then adapts this to fit in with his own knowledge of his students.

The following pages then offer multiple approaches to solving the problem introduced by cartoon like characters, which also includes their diagrams. All of this had been very confusing to me as, if this was the students' textbook, how did this encourage the students to "think for themselves" and "present their own solutions"?

As Professor Takahashi explained, although the students have these textbooks, they are not used in the lesson and students do not "look ahead" to the next lesson by turning over the page. The current series of textbooks, however, was specifically designed to encourage students to study on their own (Shimizu & Watanabe, 2010), and to enable students to relive the latest lesson at home at their own pace. Each student is given a new textbook at the start of the school year, which they are allowed to keep. Consequently, many homes have complete sets of books on their bookshelves.

I now understand how the textbook is designed to both assist teachers in their instruction via problem solving and to help students to learn through problem solving. It now makes sense. It also makes sense that textbooks in Japan have to be approved by the government to ensure they capture the essence of the current COS.

Unfortunately, there are no such textbooks in the United Kingdom but this trip has inspired me, in conjunction with Professor Burghes, to rewrite our current MEP textbooks in a similar style. In other words, write textbooks based on ideas developed through lesson study that will serve both as a teacher's guide to teaching through problem solving and a student's guide to learning through problem solving.

I was already aware that Japanese problem-solving lessons could be described as "structured problem solving". In each lesson there is one very clear learning objective based on the COS that emerges during the lesson. Teachers are rarely, if ever, side-tracked from achieving their lesson goal. I had probably not appreciated how carefully the learning objectives as laid down in the official COS are then replicated in the textbook.

As the COS is a weighty volume many teachers use the textbook to plan a series of lessons where the "learning residue" (Hiebert al., 1996) is clear and carried forward and reinforced by using it in the next lesson. We saw many examples of lessons that made reference to and built on previously learnt concepts. In the United Kingdom, we have a history of using more open problems where the endpoint is not clearly defined and not everyone will reach the same point. This possibly encourages more creativity in UK classrooms but sacrifices the more disciplined approach very apparent in the Japanese classrooms. In Japan there was more emphasis on the method (process) being open, meaning there is more emphasis on looking for different ways of solving the problem.

Students' notebooks

In my own school, we generally start the school year with high hopes that our students will use their notebooks productively. Initially we try to get our students to take pride in their work and follow basic rules in setting out their notebooks. I had seen in the Mathematics International textbook series that there is a much greater emphasis on a standard approach to how these notebooks should be set out.

Students are expected to record the date, today's problem, the student's own idea, their friend's idea, a summary of their learning and a reflection. The reflection itself covers various questions such as, "What you've come to understand", "What have you noticed?", "What do you want to examine next?" and "What you thought as you listened to your friends' idea". To remind students what is expected, each textbook contains an example of how to record their "record of learning" in one of the early chapters.

Well, that's the theory, so what did we see in the lessons we observed in Japan?

I am pleased to say we saw exactly that and yet, some of it was not quite what I anticipated. Students in the lessons we observed tended to work alone on the problem for around 10–15 minutes. There was virtually no interaction between students sitting next to each other. In the United Kingdom, we generally encourage students to try the problem for themselves for a few minutes before turning to their neighbour and sharing their initial ideas. Problems are frequently tackled in pairs with different pairs offering different solutions. It is rare for the pair to offer individual solutions as they generally reach a consensus before offering their solution to their teacher. I had assumed that Japanese students worked alone and then shared solutions with their friend. I had assumed they then recorded their friend's solution in their notebooks. This is not the case.

During the "kikan-shido" phase of the lesson, the teacher circulates around the room noting which methods of solution each student has attempted and deciding whom he will invite to present their solution during the "neriage" phase. It is only after the whole class discussion ends that the students choose which of their friends' solutions they will record in their books. It is clear that, in Japanese classrooms, everyone is regarded as a friend and so the students are free to record the solution that interested them most.

It is also clear that Japanese students take immense pride in their notebooks and all of the books I was able to see were immaculate. The use of notebooks is consistent across all years and the students understand exactly what is expected of them.

Takahashi and McDougal (2016) state that students in the United States would benefit by learning how to record the "thinking of other students. And, if students could learn to record the thinking of other students, this would be a step toward thinking critically about others' reasoning". For this to happen, teachers should model what is expected with their own board work.

Lesson planning

Fujii (2016) indicates the second step of the lesson study cycle involves developing a series of lessons that address the research theme. He also adds that one lesson in the series is then chosen as the research lesson and is planned in much greater detail by the group.

Fujii goes on:

> While the importance of a lesson plan as a product of lesson study is certainly understood, compared to the research lesson, of which there are many public examples, the collaborative work of Japanese teachers in creating a lesson plan is generally mysterious, because it is difficult to observe.

Takahashi (2014) reveals that having decided on the topic, the lesson study group then conduct *kyozaikenkyu* related to the topic. He identifies *kyozaikenkyu*, the careful study of teaching materials, as being one of the elements of lesson study that has not been successfully understood outside of Japan. It is through this process that "teachers develop their understanding of the content they teach". He adds "an important long-term outcome of the school research project would be that teachers would conduct rich *kyozaikenkyu* when preparing their everyday lessons". In this way, taking part in lesson study intentionally leads to better teaching in all lessons.

In a discussion at basecamp, Professor Fujii indicated that there are six approved textbook series in Japan and that school administrators should ensure that the norm of *kyozaikenkyu* is studying how each of these books addresses a particular topic. He added that for some topics "that have been fully researched" these approaches may be very similar but where there are clear differences these "have not yet been fully research". This once again highlighted the fact that textbooks are a product of lesson study.

Fujii (2016) explains that although much has been written about the Japanese use of structured problem solving (Becker, 1990; Doig et al., 2011; Stigler & Hiebert, 1999), the thinking behind the choice of task remains a mystery to many non-Japanese teachers. He emphasises that a major part of the planning process is discussing the "appropriateness of the task". He points out that the chosen task should be "understandable by the students with minimal teacher intervention; it should be solvable by at least some students (but not too quickly), and it should lend itself to multiple strategies". Fujii emphasises that the problem must offer plausible alternative solution methods to enable the third stage of the lesson, the whole class discussion, to be useful. This process of "polishing" students' ideas through a whole class discussion (*neriage*) is regarded as crucial to the effectiveness of the whole lesson. He also explains that in the fourth phase, *matome*, the teacher must be able to speak confidently about each strategy offered and, where appropriate, explain which method is the "most sophisticated and why". He adds that teacher also needs to fully understand the "mathematical and educational values of the task".

Use of the board: Bansho

On both of my earlier visits I had been particularly impressed with the way the teacher used the blackboard to develop the story of the lesson. "Bansho" can be translated as "use or organisation of the blackboard" or more literally "board writing". Japanese teachers consider the blackboard an important instructional tool for organising students' thoughts and therefore they take bansho very seriously (Takahashi and McDougal, 2015).

Takahashi and McDougal (2015) suggest that lesson study should be used not only develop classroom practice but also the use of the board. They go on to recommend that a goal of lesson study should be to improve students' note taking by connecting it to teachers' use of the board. They also note that in Japan, teachers use the board strategically to display different student solutions as a way to support discussion about those solutions. It is through these different approaches that the teacher can then orchestrate the discussion phase of the lesson and help the students develop their ability to think and express mathematically.

Lessons

Whilst we were in Japan, we had to opportunity to watch seven lessons. In each lesson we were specifically asked to look carefully at the four stages of the lesson and record our thinking each day in our reflective journals. These were then submitted on a daily basis to Christine Lee, the programme evaluator for IMPULS.

We were specifically asked to think about

1 Strategies to build interest or connect to prior knowledge and, in particular, how the problem was introduced.
2 Independent problem solving. Did the students work individually, in pairs, in groups, or a combination of strategies? What was the experience of diverse learners? What was the teaching doing?
3 Presentation of Students' Thinking, Class Discussion
4 Summary/Consolidation of Knowledge and in particular, strategies to support consolidation, for example, blackboard writing, class discussion, math journals.

In addition, were asked to report on three other questions

1 What new insights did you gain about mathematics or pedagogy from the debriefing and group discussion of the lesson?
2 What new insights did you gain about how administrators can support teachers to do lesson study?
3 How does this lesson contribute to our understanding of high-impact practices?

For reporting purposes, we were split into groups of four teachers with each group being asked to report on one specific lesson.

More detailed look at a lesson

My group were asked to comment on a Grade 6 lesson entitled. "What is the size of the shaded part?" We had already been given a detailed lesson plan.

The stated goals of the unit were that

- Students will be able to calculate the area of figures.
- Students will be able to approximate figures in their surroundings and determine their area.

The lesson plan or lesson proposal (Fujii, 2016) outlines the related learning that has taken place in previous years and points out the students first met π in Grade 5 as the relationship between the diameter and circumference of a circle.

In lesson 4 of this unit the students had studied ways of calculating the area of a circle and developed the area formula. The goal of the lesson was that students could think about a variety of strategies to determine the area of a composite figure involving circles and calculate the area. This is lesson number 5 of 10.

The lesson plan also includes specific strategies that will be used in today's lesson that address the research theme. These strategies link directly to the phases of a problem-solving lesson described earlier, as can be seen in the following text:

1 Engaging students with the task and drawing out the question (Grasp).
2 Experiencing the joy of solving a problem independently through solving a genuine problem (Explore).
3 Through the activity of interpreting other students' ideas, students will answer their questions and deepen their reasoning (Deepen/Heighten).
4 Through the activity of summarising, applying, and extending, draw out additional questions from students (Summarise/Extend).

The notes that follow are taken from the group report together with my own observations.

Introduction – posing the task – [Strategies to build interest or connect to prior knowledge and, in particular, how the problem was introduced. Engaging students with the task and drawing out the question (Grasp)].

The lesson began with Mr N welcoming the class and the observers to today's lesson. The students are clearly at ease with Mr N and unfazed by the number of other people in the room. The students' desks are completely clear at this point.

Mr N then produced the "mystery box" which, in itself, quickly captured the students' attention and created a sense of intrigue. All eyes focus on the mystery box as Mr N slowly draws the drawing of the first shape from the mystery box. As he does so, he asks, "What is the area of the coloured part?"

The first shape revealed is a circle, radius 10 cm that is then stuck to the left-hand side of the board. There is a general murmur of recognition and after a few moments Mr N the selects one student to explain how to calculate the area of a circle that had been covered in the previous lesson. Mr N listens carefully and writes underneath the picture,

$$\text{Area} = 10 \times 10 \times 3.14 = 314 \text{ cm}^2.$$

He then slowly draws out the second shape that is a semicircle and the process is repeated with the shape being stuck alongside the first shape on the left-hand side of the board. Mr N compliments the students on their answers so far.

He then slowly draws out the third shape that the students were anticipating would be a quarter circle. It isn't and there is a buzz around the room as the students see an unfamiliar shape.

Mr N then asks the students why they were surprised. One of the students explains that he thought it would be a quarter circle. Mr N then asks, "Could you find the area if it were a quarter circle?" There is general agreement that this would be easy. When the shape was revealed, Mr N asked the students to name the shape and what shapes they could see. The students named the shape 'Lemon'. This was different from what the teacher had anticipated (Leaf).

He then added "Let's think about ways of calculating the area of the lemon shape". The lemon shape is then stuck beneath the first two shapes on the left-hand side of the board.

Mr N then wrote today's problem in the centre of the board and as he did this he asked the students to take out their notebooks.

"Let's think about ways to calculate the area of the coloured part (the lemon)". He stuck a second copy of the shape to the right of the problem.

A student was then selected to read the question and then the students wrote the question into their notebooks on a new page. It is clear that the students all know their roles in this type of lesson (in contrast to the United Kingdom).

"Check – does everyone understand the question?" (Student reads out the question.)

All of the above was done with pace and was completed in less than 5 minutes.

The introduction met the first point of observation which was to capture the students' imagination and motivate them to engage in the task.

Independent problem solving – [Did the students work individually, in pairs, in group, or combination of strategies? What was the experience of diverse learners? What was the teaching doing? Experiencing the joy of solving a problem independently through solving a genuine problem (Explore).]

Mr N then gave everyone a copy of the diagram to help his or her thinking but explained that it was smaller than the actual size. He tells his students to "use words and diagrams to explain your ideas".

"I am going to give you 3 minutes to start". But in reality, he only gives them 1 minute "If you are stuck, please come to the front of the room (referred to as his maths lab)".

A small group of 10 students came to the front of the room where Mr N helped by asking specific questions and using a selection of cut out shapes to guide their thinking.

These pupils, however, were then directed towards a specific strategy but it did mean they were able to start the question. The intervention, though, used though provoking questions to stimulate the students thinking. As each student began to understand the problem, they slowly moved away to work independently.

Interestingly only 5 of the students did actually use this strategy and the other 5 came up with other ideas when they returned to their seats.

During this phase of the lesson the students began to record their ideas in their notebooks using mathematical expressions, words and diagrams. There is no discussion with their neighbours – it is an individual effort with students clearly taking pride in the presentation of their work.

Mr N meanwhile moved around the room making careful notes of the various strategies being used - and who was using them. During this phase of the lesson, he decides which students will be asked to present their ideas in phase 3 of the lesson. In the lesson plan, the planning team had identified three different approaches that they thought might arise. Interestingly, all three approaches identified in the lesson plan were used by the students.

It was clear that some students finished very quickly; they then spent time writing their explanation in great detail or waiting patiently.

Presentation of Students' Thinking, Class Discussion [Covering student thinking/visuals/peer Responses/Teacher Responses. Through the activity of interpreting other students' ideas, students will answer their questions and deepen their reasoning (Deepen/Heighten).]

I know some of you are still working but we now need to begin the discussion phase (again all students know their role and so there is no dissent (contrast to United Kingdom). Mr N now invites a pre-selected student to explain their method of calculation, from their seat. Mr N writes the calculation in centre section of the board.

He then asks a second student to come to the board and stick the correct shapes on the calculation, so that it is clear where the calculation has come from. Mr N now draws a box around this solution.

Mr N checks to make sure that everyone understands.

Another student is then asked to give her equation from her seat without any explanation. Mr N again writes the solution on the board.

This time a different student explains what the equation means (no apparent resentment from the student – this is clearly normal practice). Finally, another student adds the shapes to parts of the equation.

There is a very clear emphasis on interpreting other students' ideas throughout, thus building a shared understanding of the strategy.

The teacher now asks another student to explain his method. This follows a similar pattern to the above with Mr N writing the equation on the board and then other students interpreting parts of the equation and sticking the shapes into place.

Throughout this phase Mr N was very patient with the students as they presented and explained their answers.

All of Mr N's operational commands were given in an affirmative supportive manner. Throughout the lesson precise language is used when describing the triangle as an isosceles right-angled triangle.

Summary/Consolidation of Knowledge – [Strategies to support consolidation, e.g., blackboard writing, class discussion, math journals. Through the activity of summarising, applying, and extending, draw out additional questions from students (Summarise/Extend).]

In summarising the lesson, Mr N emphasised that they were able to find the unknown area using shapes they already know how to calculate the area. He specifically asks them to think about what kind of shapes they used to solve the problem.

In this phase of the lesson the students are encouraged to write down their favourite solution from what they have seen from the work presented on the board or from their neighbour. Mr N wanted the students to explain why they chose a particular method but the lesson is running late and there is no time for this to take place.

Most simply copy off the board.

The students are very proud of their notebooks and do not like mistakes. If they make any, they rub them out carefully. Their aim seems to be to produce a perfect book whereas keeping a record of their mistakes might actually be useful.

Teacher ends the lesson by returning to the mystery box where he reveals three more puzzles for tomorrow's lesson. There is general excitement and both teacher and students appear to be looking forward to tomorrow's lesson.

Post lesson discussion

The planning team speak first explaining the reasons for choosing to deliver this lesson which is very much as in the textbook but with a modified hint. The thinking behind this is explained.

The teacher then speaks, explaining how he implemented the lesson plan in the lesson. He explains that he is pleased that the students came up with the three ideas that the team had suggested in their planning. He acknowledges that he ran out of time and so cut short the summary phase of the lesson. Others then speak giving suggestions and asking questions.

Finally, the visiting speaker sums up and seems to include more general comments about maths education than about the lesson. He does however stress the importance of linking the lesson to the COS. He also emphasises that the summing up phase of the lesson ("matome") should respond to the lesson objectives as well as the lesson (often forgotten in a busy lesson). He also points out that notetaking should be about the recording of pupils' own ideas, not the collective work of the group.

Insights gained from the debriefing and group discussion of the lesson

The link between mathematical modelling and physical images that are used to build collective understanding are very skilfully demonstrated in the lesson. In the United Kingdom, we have a linear hierarchy of moving from "concrete" to abstract. Maybe this is something we need to reconsider and develop students' abilities to "create first in the mind".

The balance between "teacher speak" and "student speak": the teacher primarily led the discussion, with student comments usually being directed back to the teacher. It would be interesting to know the level of understanding of those students that did not speak in the lesson. Furthermore, since the majority of the students will have the same notes as the teacher, how will an analysis of these notes enable the teacher to establish the level of understanding of the lesson?

The objective of the lesson was achieved as stated in the lesson plan, that is, the summary of the learning was as stated. The question is whether this was accurate in terms of being a true reflection of the learning or whether the teacher was just able to achieve this as a result of the collective work on the board.

Insights gained about how administrators can support teachers to do lesson study

The leadership of schools needs to be committed to embedding the process into their professional development programs. The culture of "teacher as researcher" needs to be embedded. In the United Kingdom, we do have the time to do this, but the challenge would be for the headteacher to have the confidence to reduce the intensity of learning walks and lesson observations in order to create capacity elsewhere. This would require a shift in culture that has been developed in the United Kingdom through a rigorous and punitive accountability regime.

The lesson study approach used in Japan focuses on the deepening of subject knowledge and the nurturing of engagement in learning. In the United Kingdom, we have been exposed to a range of teaching and learning strategies all of which have systematically failed to embed a culture of teacher as learner and the development of their own craft.

The lesson study in the school is a whole school activity with an overarching aim. The research question has been thoughtfully developed in consideration of the whole school priority and is worked on for an extended period of time (two years is not unusual).

In addition to these points, as the United Kingdom is currently looking at ways to adopt a 'mastery' style to teaching and learning, and programs of learning and assessments are being reformed at every stage of education to introduce new content and make the curriculum more challenging, there is heightened attention for CPD that improves subject knowledge and improving ways of introducing content to students. This is in contrast to the type of CPD that has previously been popular in the United Kingdom focusing on teaching style (e.g., three-part lesson).

Therefore, although the time is right to implement lesson study to the United Kingdom and indeed in developing countries, we anticipate that it could cause some tension.

The detailed planning and importantly the consideration of anticipated student responses is a key feature of high impact practice. This process not only enables the teacher to manage the 'flow' of the lesson but also deepens the teacher's own subject knowledge and "teaching craft".

The development of an affirmative learning culture is also a key area of high impact practice. It is clear that, over time, students are nurtured to take responsibility for their own learning and as the students mature, they increase their engagement in the lesson not necessarily by a direct contribution, but by listening carefully to the teacher and others and by detailed and effective note taking.

The culture of the "productive struggle" is also an important feature. Students are asked to consider the problem on their own, and only if they cannot get started do they ask for help. This builds up resilience and confidence in approaching new problems and mathematical situations.

Further thoughts: Choice of task

This task comes directly from the textbook. In many ways this was a surprise to me as in the United Kingdom we generally try to think of a completely new lesson in our lesson study cycles. This then reinforces the comments written earlier about how teachers use the textbook to plan their lessons. In our group discussion later at Basecamp, Professor Fujii explained that textbooks are largely the product of lesson study and that by starting with an idea from the textbook, the teacher is building on the work of others rather than beginning again.

The task clearly met Fujii's expectations of a suitable task as

- it was understandable by the students with minimum intervention;
- it was solvable by most if not all students (although possibly solved too quickly by some);

- it offered multiple solutions which the teacher was able to elicit from his students;
- this then enabled a productive whole class discussion.

Further thoughts: Use of the board

It was obvious from this lesson that the success of the third and fourth phases of the lesson not only depended on the choice of task but also on way the teacher used the board. The advantages of having a large board were obvious, which poses the question of "How can we mirror the use of the board when most schools in the United Kingdom have abandoned chalk boards for smaller interactive boards?"

Implementation

As explained elsewhere, lesson study is based on a research theme or overarching aim. As covered by Burghes and Robinson (2010), the overarching aim of the mathematics department at my school is currently

"Through our mathematics teaching our students will become independent thinkers (learners) who enjoy working together to produce creative solutions in unfamiliar situations".

Following Killion (2008), we then turned the overarching aim into specific, measurable objectives that are used to evaluate the success of our research lessons over the next three years:

1 Enjoy doing mathematics – to help students learn to enjoy and sense personal reward in the process of thinking, searching for patterns and solving problems.
2 Gain confidence and belief in their abilities – to develop students' confidence in their ability to do mathematics and to confront unfamiliar tasks
3 Be willing to take risks and to persevere – to improve students' willingness to attempt unfamiliar problems and to develop perseverance in solving problems without being discouraged by initial setbacks
4 Interact with others to develop new ideas – to encourage students to share ideas and results, compare and evaluate strategies, challenge results, determine the validity of answers and negotiate ideas on which they all can agree.

Building on the evidence from a comprehensive literature review and the experiences gained from my time in Japan, I now felt able to suggest certain design principles that could be applied in England and other countries, including developing countries, that would capture the Japanese approach to teaching through problem solving and could form the basis of "great instruction".

1 Students do not learn passively but learn through mathematical activities where "mathematical activities" mean various activities in which students willingly and purposefully work on mathematics. They specifically exclude activities where students just listen to teachers' explanations or complete practice problems that are not included in mathematical activities.
2 Mathematics is a body of well-connected ideas based around big ideas or through lines that are central to a topic and often span many different school years. Consequently, successful units of lessons make these through lines explicit together with clear unit goals and specific lesson objectives that enhance students' learning. These should be defined in changes of understanding that will occur during the lesson.
3 Successful problem-solving lessons are not stand-alone lessons. They build on prior knowledge and drive the students' learning forward in small steps. In general, they will be an essential part of a series of lessons. Consequently, it is important that the chosen activity links to previous and future ideas. It is important that both the students and the teachers know what has gone before and what is to come. This implies that we need a very detailed scheme of work similar to the Japanese COS.
4 Activities must give the students opportunities and ample time to develop their ability to think, explain and express their thoughts mathematically (COS).
5 Problems should be chosen that capture the students' attention so that the students generate questions and solutions of their own. The problem should be presented in a way that captives the students through their sense of curiosity or through surprises.
6 The problem should be chosen so that all students are able to arrive at a solution with the minimum of help from the teacher. Thought should be given to the provision of necessary learning aids that might be required by some students to enable them to accomplish the learning objectives.
7 The problem should generate different solution methods from the students, enabling them to take ownership of solving the problem. Deep learning occurs when students compare and discuss the different solution methods.

Towards the end of the summer term of 2016, both of my lesson study groups met to hear what I had learned from my trip to Japan. I outlined much of what is reported above and suggested that everyone included "*kyozaikenkyu*" in future lesson study cycles.

I also introduced the tentative design principles and, in the discussions that followed, there was general agreement that Principles 1–3 were already well understood by the lesson study teams. It was also noted that we needed to think about how we might incorporate Principles 4–7 in future cycles. I then explained that in Japan, the main learning takes place in the *"neriage"* phase of the lesson where students have the opportunity to compare and contrast several different solution methods that the students themselves devised during the *"kikanjunshi"* phase. This then led onto an in-depth discussion about the significance of Principle 7 namely that "the problem should generate different solution methods from the students". It was agreed that one of the teams of teachers would plan a research lesson (or prototype) that included all of these principles in the next cycle but with particular emphasis on Principle 7.

We had previously identified algebra as an on-going difficulty for many students throughout the school. As Charge (2017) explains

> Although many can manipulate algebraic expressions accurately, they see little point in it and we are constantly asked, "Why am I doing this?" and "When will I ever use this?" For many students, the introduction of letters to represent unknown quantities remains a mystery.

Consequently, at their first meeting of the school year, one of the school's lesson study teams decided to make "Developing algebraic thinking" the theme for their next lesson study cycle. At this meeting, the team leader reminded everyone of the importance of *"kyozaikenkyu"* (study of teaching materials) and asked the members of the team to find at least one way that algebra is currently introduced into schools.

The Team Leader explained they should look in textbooks, online and through their own resources and present their ideas at the next meeting. The details of the *"kyozaikenkyu"* are reported by Charge (2017).

As Takahashi (2016) had explained, many research lessons use ideas taken from the textbook and modified to fit a particular group within a school. Consequently, the team leader intentionally looked at how the Japanese introduced algebra in Grade 7 and thought about how to adopt this for use in this unit.

At the next meeting the Team Leader introduced a problem that he thought would lead to the students themselves choosing to use letters to represent "any number". During the *"kyozaikenkyu"* phase it had been noted that at Key Stage 2 (age 9–11 years), students were introduced to algebra through "finding an unknown" which was sometimes represented by a shape, a symbol and later by a letter. He outlined that we wanted the students to use letters for variables, a new concept for all of our students.

So we found a starting point for our cycle of lesson study but it perhaps needed to be adapted and revised in order to ensure that it would meet Principle 7 as well as the known issues we have in teaching Algebra to classes throughout the school.

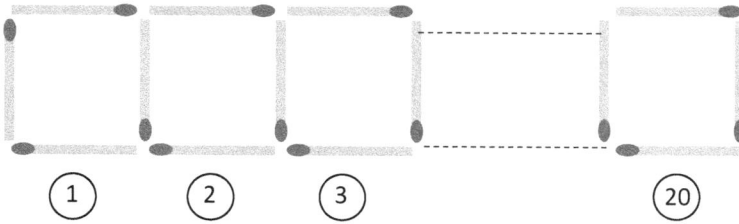

Figure 11.1 Diagram for matchstick problem posed in the Japanese text book.

Figure 11.1 shows the diagram from the textbook page; students were asked how many matchsticks would be needed to make 20 squares in a similar way.

At our next meeting, the group began planning a series of lessons that they thought would address both the research theme (overarching aim) and the unit goals, which were:

- Students will discover their own rules for growing patterns;
- They will generalise these patterns in words;
- They will begin to generalise using algebraic methods.

Further details are explained by Charge (2017) but note that the matchstick problem was modified to:

A Square Problem

"Squares are made using matchsticks as shown below.

How many different ways can you find of counting the matchsticks needed for 5 squares?"

The series of lessons designed is summarised as follows.

The detailed review of our implementation is given by Charge (2017) but, to summarise our detailed discussions that followed, the group agreed that in the series of lessons based around our matchstick problem, we had met all of the principles.

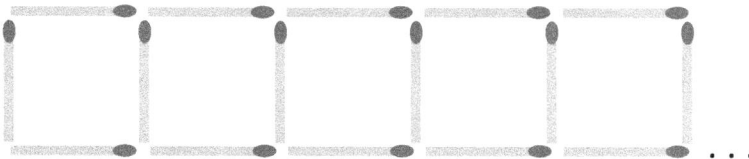

Figure 11.2 The modified problem presented.

Table 11.1 Sequence of lessons designed to meet the principles outlined earlier

	Learning activity
Lesson 1	**Are you ready?** Students will be asked to generate questions of their own from real world situations.
Lesson 2	**Growing patterns.** Students will generalise a growing pattern in words. Intentionally this use of words makes this cumbersome – but this is revisited in lesson 4.
Lesson 3	**Consecutive numbers.** Students will work on a number puzzle, initially with small numbers but later with large numbers that simply don't fit into the box. What will they decide to do? The intention is that they choose to use m to represent a million and later n to represent any number. This is a research lesson.
Lesson 4	**Growing patterns revisited.** At the end of the previous lesson students were challenged to try to generalise their growing patterns from lesson 2 using n to stand for any number. The intention is that they realise using letters is much simpler than using words.
Lesson 5	**How many matches?** Although this is a familiar task the intention here is to implement some of design principles suggested earlier. This is also a research lesson and is described in Charge (2017).
Lesson 6	**What have I learnt?** A final lesson in this sequence of lessons which gives the students a chance to demonstrate their ability to generalise succinctly.

Lessons learnt

In Japan, teachers learn their craft by taking an active role in lesson study throughout their career. It is during these cycles that novice teachers rub shoulders with expert teachers and learn the art of teaching through problem solving. It is also through lesson study that new curriculum reforms are introduced over a 3-year period based on a meticulously constructed COS. This COS makes the design principles behind the changes very clear and stresses that it is through activities that students learn mathematics as highlighted. Textbooks are then complied from the information drawn from lesson study by practising teachers and educational experts. Textbooks are used quite differently in Japan as a planning tool for teachers and a reinforcing tool for students working at home. All students are given copies of all of their textbooks to keep forever and use them for reference in later years.

Unfortunately, none of these things are currently in place in England or developing countries, where teachers are often given little guidance on how to introduce change and almost no time to plan for implementation. Consequently, in England, for example, textbooks are then rushed out, usually produced by examination boards and designed for delivering their exam syllabus. There appears to be a belief at the highest level that change can happen instantaneously. This view is clearly not shared with other, more successful, countries.

This needs to change if we are going to produce a problem-solving curriculum in Mathematics that motivates both the teachers and their students.

Implementation in developing countries

Japanese educators (Fugii, 2014) warn that countries intending to introduce Japanese lesson study as their main form of professional development need to "grasp its fundamental nature and describe it carefully" if it is to be successful. As Stigler and Hiebert stated earlier, teaching is a cultural activity and changing cultural scripts is very difficult.

Schools in developing countries will need to import expertise from elsewhere, preferably from Japan, if they are to appreciate the nuances of lesson study that Japanese teachers take for granted. In a similar way, teaching through problem solving is not as straightforward as it sounds and again, developing countries may need to import expertise.

This report indicates that the Japanese approach to teaching through problem solving could be implemented effectively in schools throughout the world, given time, money and expertise. It does, however, require more research before any greater claims can be made.

There are many open questions about its wider use that could form the basis of future research and development. Some possible questions could include:

*   Will teaching through problem-solving work in more challenging schools?
*   Can teaching through problem solving be implemented in schools without an expert (knowledgeable other) to offer support and influence?
*   Can teaching through problem solving be introduced piecemeal or does it need to be a whole school or country initiative?
*   Can teaching through problem solving succeed in schools that rely on more traditional teaching approaches?

Finally though, I would like to encourage schools and teachers, despite the caveats above, to consider how they can undertake lesson study in their school, primary or secondary, as it is an immensely powerful way of implementing change and bringing lessons to life with both teachers and students looking forward to their next lesson.

References

APEC Education Network (EDNET). (2003). Guide for Planning and Analyzing Mathematics Lessons in Lesson Study. [Online]. Available at: http://hrd.apecwiki.org/index.php/Guide_for_planning_and_analyzing_mathematics_lessons_in_lesson_study. Accessed on January 10, 2013.

Bartley, K. (23 Nov. 2007). What makes for effective teaching? *Standing Conference on teacher education North and South (SCoTENS) entitled 'Teaching in the knowledge society'*. County Dublin, Malahide.

Becker, J. P. (1990). Some observations of mathematics teaching in Japanese elementary and junior High schools. *Arithmetic Teacher*, *38*(2), 12–21.

Becker, J. & Epstein, J. (2005). Reflections on U.S. Collaborative Research in Mathematics Education. In: *U.S. – Japan Collaboration in Mathematics, Science and Technology Education*. National Science Foundation, USA.

Becker, J. P. & Shimada, S. (Eds.) (1997). *The Open-Ended Approach: A New Proposal for Teaching Mathematics*, NCTM, VA.

Burghes, D., & Robinson, D. (2010). *Lesson study: Enhancing mathematics teaching and learning*. CfBT.

Cambridge Assessment. (2012). Exams should not define the curriculum: Achieve – Spring 2012. [Online]. Available at: http://www.cambridgeassessment.org.uk/Images/140056-achieve-spring-12.pdf. Retrieved April 10, 2017.

Charge, S. (2017). *Developing algebraic thinking through Japanese lesson study*. Plymouth University.

Cockcroft, W. H. (1982). *Mathematics counts: Report of the committee of inquiry into the teaching of mathematics in schools*. H.M. Stationery Office.

DfE (2013). Mathematics programmes of study: key stage 3 National curriculum in England. DFE-00179-2013.

Doig, B., Groves, S., & Fujii, T. (2011). The critical role of task development in lesson study. In L. Hart, A. Alston, & A. Murata (Eds.), *Lesson study research and practice in mathematics education* (pp. 181–199). Dordrecht, The Netherlands, Springer.

Fernandez, C., & Yoshida., M. (2004). *Lesson study: A Japanese approach to improving mathematics teaching and learning*. Mahwah, NJ, Lawrence Erlbaum Associates.

Fujii, T. (2013). Implementing Japanese Lesson Study in Foreign Countries: Misconceptions Revealed.

Fujii, T. (2014). Implementing Japanese lesson study in foreign countries: Misconceptions revealed. *Mathematics Teacher Education and Development, 16*(1), 65–83.

Fujii, T. (2016). Designing and adapting tasks in lesson planning: A critical process of lesson study. *ZDM, 48*(4), 411–423.

Guskey, T. (2000). *Evaluating Professional Development*, Corwin Press Inc. California.

Hiebert, J., Carpenter, T. P., Fennema, E., Fuson, K., Human, P., Murray, H., Olivier, A., & Wearne, D. (1996). Problem solving as a basis for reform in curriculum and instruction: The case of mathematics. *Educational researcher, 25*(4), 12–21. https://doi.org/10.2307/1176776

Hopkins, D. (2007). *Every school a Great school*, McGraw-Hill Education, Open University Press.

IMPULS. (2016). International math-teacher professionalization using lesson study. Available at: http://www.impuls-tgu.org/en/. Retrieved April 11, 2017.

Isoda, M. (2007). Where did lesson study begin, and how far has it come?. In M. Isoda, Y. Ohara, T. Miyakawa, & M. Stephens (Eds.), *Japanese lesson study in mathematics: Its impact, diversity and potential for educational improvement* (pp. 8–15). New Jersey, World Scientific Publishing Co. Pte. Ltd.

Isoda, M., Stephens, M., Ohara, Y., & Mijakawa, T. (2007). *Japanese lesson study in mathematics*. World Scientific Publishing.

Isoda, M., & Katagiri, S. (2012). *Mathematical thinking: How to develop it in the classroom*. World Scientific Publishing.

Japanese Ministry of Education. (2008). Available online at: http://www.globaledresources.com/products/books/guide_arithmetic_7-9.html

Killion, J. (2008). *Assessing Impact*, Corwin Press, California.

Kishimoto, T., & Tsubota, K. (2007). What are the features of lesson study projects conducted in elementary school mathematics departments? In: M. Isoda, Y. Ohara, T. Miyakawa, &

M. Stephens (Eds.), *Japanese lesson study in mathematics: Its impact, diversity and potential for educational krathwohl, D.R., 1993. Methods of educational and social science research: An integrated approach.* Longman/Addison Wesley Longman.

Lewis, C. (2002). *Lesson study: A handbook of teacher-led instructional change.* Philadelphia, Research for Better Schools.

Lewis, C. (2003). The essential elements of lesson study. *Northwest Teacher*, vol. 4, no. 4 (Spring 2003), 6-8. Becker (2005).

National Research Council. (1989). *Everybody Counts: A Report to the Nation on the Future of Mathematics Education.* Washington, DC: The National Academies Press. https://doi.org/10.17226/1199.

Natsusaka, S. (2007). Discussion-orientated teaching methods and examples: The problem-solving orientated teaching methods and examples. In M. Isoda, Y. Ohara, T. Miyakawa, & M. Stephens (Eds.), *Japanese lesson study in mathematics: Its impact, diversity and potential for educational improvement* (pp. 92–101). New Jersey, World Scientific Publishing Co. Pte. Ltd.

National Council of Teachers of Mathematics (NCTM). (1980). *An agenda for action: Recommendations for school mathematics of the 1980s.* Reston, Virginia, NCTM.

National Council of Teachers of Mathematics (NCTM). (1989). *Curriculum and evaluation standards for school mathematics.* Reston, Virginia, NCTM.

Ofsted. (2011). Key messages from Ofted's recent mathematics reports. Available at: https://www.whatdotheyknow.com/request/154950/response/384078/attach/16/Mathematics%20reports%20distance%20learning%201.pdf. Retrieved April 11, 2017.

Pehkonen, E. (2008). Problem solving in mathematics education in Finland, p. 3. Available at: www.unige.ch/math/EnsMath/Rome2008/WG2/Papers/PEHKON.pdf

Polya, G. (1945). *How to solve it: A New aspect of mathematical method.* Princeton, NJ, Princeton University Press.

Schoenfeld, A. H. (1992). Learning to think mathematically: Problem solving, metacognition, and sense-making in mathematics. In D. Grouws (Ed.), *Handbook for research on mathematics teaching and learning* (pp. 334–370). New York, MacMillan.

Schroeder, T. L., & Lester, F. K., Jr. (1989). *Developing Understanding in mathematics via problem solving.* In P.R. Trafton [Ed] New directions for elementary school mathematics, 1989 Yearbook of the National Council of Teachers of Mathematics (pp 31–42) Reston, V. A: NCTM.

Shimada, S. (1997). The Significance of an Open-Ended Approach. In: P. Becker & S. Shimada (eds.), *The Open-Ended Approach: A New Proposal for Teaching Mathematics,* NCTM, VA.

Shimizu, Y. (2007). Understanding Japanese mathematics lessons section 6.1: How do Japanese teachers explain and structuralize their lessons?. *Japanese Lesson Study in Mathematics: Its Impact, Diversity and Potential for Educational Improvement* (p. 64). World Scientific Publishing Co Pte Ltd.

Shimizu, S., & Watanabe, T. (2010, March). Principles and processes for publishing textbooks and alignment with standards: A case in Japan. In *APEC conference on replicating exemplary practices in mathematics education, Thailand* (pp. 7–12).

Skemp, R. R. (1976). Relational understanding and instrumental understanding. *Mathematics Teaching,* 77(1), 20–26.

Singapore Ministry of Education. (2007). Available online at: http://www.moe.gov.sg/education/syllabuses/sciences/files/maths-primary-2007.pdf

Stigler, J. (2000). *Before it's too late: A report to the nation from the National Commission on Mathematics and Science teaching for the 21st Century.* Diane Pub Co.

Stigler, J., & Hiebert, J. (1999). *The teaching gap: Best ideas from the World's teachers for improving education*. New York, Free Press.

Sugiyama, Y. (2008). Introduction to elementary mathematics education (in Japanese). Tokyo: Toyokan Publishing Co.

Takahashi, A. (2011). The Japanese approach to developing expertise in using the textbook to teach mathematics rather than teaching the textbook. In *Y. Li & G. Kaiser (Eds.), Expertise in mathematics instruction: An international perspective* (pp. 197–219). Dordrecht, The Netherlands, Springer

Takahashi, A. (2014). Supporting the effective implementation of a new mathematics curriculum: A case study of school-based lesson study at Japanese Public Elementary School. In *Mathematics Curriculum in School Education* (pp. 417–441). Dordrecht, Springer.

Takahashi, A. (2016). Recent trends in Japanese mathematics textbooks for elementary grades: Supporting teachers to teach mathematics through problem solving. *Universal Journal of Educational Research, 4*(2), 313–319.

Takahashi, A., & McDougal, T. (2016). Collaborative lesson research: Maximizing the impact of lesson study. *ZDM, 48*(4), 513–526.

Tall, D. (2008). Using Japanese Lesson Study in Teaching Mathematics. *The Scottish Mathematical Council Journal, 38*, 45–50.

Van de Walle, J. A., Karp, K. S., & Bay-Williams, J. M. (2010). *Elementary and middle school mathematics: Teaching developmentally*. Pearson.

Chapter 12

Enhancing mathematics education in developing countries

David Burghes and Naomi Sani

Setting the scene

Confidence and competence in Mathematics and Statistics of young employees are key priorities for the economies of all countries, but particularly for developing countries. With continuing developments in technology and international economic competitiveness, almost all countries are pursuing initiatives to enhance mathematics and statistics teaching and learning in schools to ensure that the next generation of young employees are able to meet the needs of business, commerce and industry. They need transferable problem-solving skills to ensure that they are confident and capable with expertise in numeracy, statistics and mathematics.

To ensure that young people to have these problem-solving skills crucially depends on how they are educated at school which leads directly to the consideration of pedagogical reforms to the teaching and learning of mathematics and related disciplines. This in turn needs teachers of mathematics with not just competent subject knowledge but an understanding of what it means to be a mathematical *thinker* as well as a problem solver with the desired transferable skills.

Here we will consider in some detail on the topic of problem solving in mathematics education as well as teaching and learning strategies to effectively implement problem solving. We will focus on suitable policies that developing countries can employ to enhance the mathematical and problem-solving skills of their future workforce. Evidence from Finland shows clearly that significant investment in education at all ages is the optimum way forward, but here we will suggest **three priority actions** that could have both short term and longer-term success.

Problem solving

There is some confusion when we use these terms but following Nohda (2000), I think that this definition is both helpful and illuminating:

A **problem** occurs when students are confronted with a task and there is no prescribed way of solving the problem.

It is generally not a problem if students can immediately solve it.

This still leaves open the question of how we should use problem solving in our teaching. For example, we can use:

- problem solving *to reinforce* a concept;
- teaching *for* problem solving;
- teaching *through* problem solving.

All these approaches have some value and, traditionally, problem-solving questions have been used to show understanding of a new mathematical concept *after* the new concept has been taught. Here we will concentrate on the 'teaching *through* problem solving' as we see this as central to learners becoming mathematical thinkers with the confidence to apply their mathematical skills in different contexts. Current mathematics education reforms both here in England and around the world suggest that "problem solving and investigative approaches are central to learning for all pupils" (Ofsted, 2011) and also see Chapter 11 in this monograph. This is certainly true in many of the countries the English government wishes to emulate.

In **Singapore**, for example, mathematical problem solving is central to mathematics learning (Singapore Ministry of Education (MOE), 2007). It involves the acquisition and application of mathematics concepts and skills in a wide range of situations, including non-routine, open-ended and real-world problems. The development of mathematical problem-solving ability is dependent on five interrelated components, namely,

Concepts, *Skills*, *Processes*, *Attitudes* and *Metacognition* (MOE, 2007).

This is illustrated in Figure 12.1.

In **Japan**, the new Course of Study (COS) outlines the importance of developing new mathematical activities that inspire the students to "willingly engage in mathematics with purpose" by "trying to find new properties or to create new ways of thinking or to solve concrete problems" (Japanese Ministry of Education, 2008).

Isoda and Katagiri (2012) summarise this by reporting that the basic principle of the problem-solving approach is to nurture children's learning of mathematics by/for themselves. The APEC guide goes on to stress that:

The problem or task you will provide in the lesson must help students develop the goal understanding. It should also capture students' attention. Thus, setting up the problem/task in the context that is meaningful to students is very important. The point is not necessarily to give students 'real-world' problems. Rather, we want to capture students' attention so that they generate their own questions.

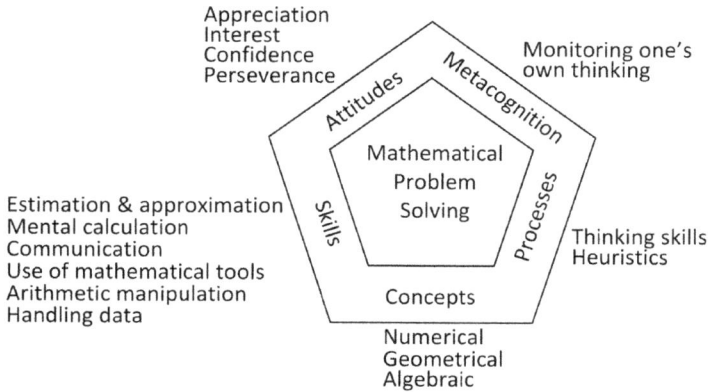

Figure 12.1 Developing mathematical problem solving in Singapore.

This is not a new idea as back in 1980 the National Council of Teachers of Mathematics (NCTM) suggested that "problem solving must be the focus of school mathematics" (NCTM, 1980, p.1). It concluded with the publication of *Everybody Counts* (National Research Council, 1989) and the *Curriculum and Evaluation Standards for School Mathematics* (NCTM, 1989), both of which emphasise problem solving.

We also have examples of the central use of problem solving in countries including Finland, New Zealand and Japan in earlier chapters of this monograph and referring to England, its inspection service (2011) certainly seem to think that problem solving lies at the heart of outstanding teaching as can be seen from their current guidelines:

- Teaching is rooted in the development of all pupils' conceptual understanding of important concepts and progression within the lesson and over time;
- It enables pupils to make connections between topics and see the "big picture";
- Teaching nurtures mathematical independence, allows time for thinking and encourages discussion;
- Problem solving, discussion and investigation are seen as integral to learning mathematics;
- Constant assessment of each pupil's understanding through questioning, listening and observing enables fine-tuning of teaching.

To clarify these ideas, we will consider a number of problems that we have used to illustrate how problem solving can be used to enhance the teaching of mathematical concepts and indeed, as in Japan, to show how problem solving can be used not just to enhance mathematical skills and knowledge but to introduce new concepts.

Examples of mathematical problem-solving activities

Historical codes: Braille

Braille is a method of representing letters and numbers, etc. which visually impaired people can read by touch.

It was designed by the Frenchman Louis Braille in 1833 and is still in use today.

It is based on using 6 dot positions in a 2 × 3 array for each letter of the alphabet. In the example shown, for the letters 'H' and 'Q', the raised dots are denoted by black circles and flat spaces by small white circles (Figure 12.2).

Problem: *How many different codes can you find?*
Extension: *Are there sufficient codes to cope with numbers, letters, punctuation, etc.?*

This problem is interesting as it can be set for very different age groups and the strategies they use to solve it can be quite varied. You can, for example, build up from

1 dot (2 possibilities) 2 dots (4 possibilities), 3 dots (8 possibilities),...
so that 6 dots give $64 = 2^5$ and this can be generalised (also reinforcing the use of indices).

Younger learners will be keen to draw out the possible codes, using 1 raised dot out of the 6 available to give 6 different codes, then using 2 raised dots out the 6, and hence build up the possible different codes in this way. This needs a logical systematic approach.

The extension problem is also interesting as a strategy is needed to cover not just the basic numbers, letters and punctuation but also capital and italic letters and much more! It works though by having a letter code and number

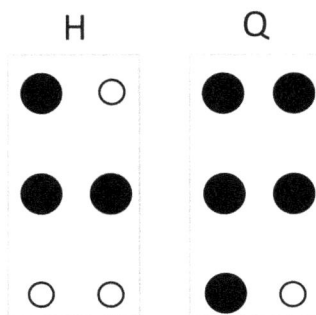

Figure 12.2 Braille dot positions for the letters H and Q.

Table 12.1 Scores from the judges in an ice-skating competition

	Judge 1	Judge 2	Judge 3	Judge 4	Judge 5
Jenna	8	6	10	9	7
Kim	9	9	7	8	7

code before strings of letters or numbers. In this way, for example, both "a" and "1" have the same code. As well as encouraging mathematical and logical thinking, it illustrates how mathematics is applied to real-world problems.

Competitive skating

In an ice-skating competition Jenna and Kim were the top two competitors. The five judges gave them the scores provided in Table 12.1.

> **Problems:** *Can you give a reason why Jenna might have been declared the winner.*
> *Why might Kim think that she should have won?*
> **Extension:** *Research the methods used in international competitions*

This is suitable for learners who have some knowledge of basic statistics. The obvious way to tackle this problem is to find the mean and median (which are identical); this leads on to the concept of consistency or variation. There is no 'right' answer; the method of determining the winner should be in place before the competition.

This extension problem leads on to a more detailed analysis of the rule used in international competitions where the highest and lowest scores from the range of the judges' scores are deleted in an attempt to avoid bias.

There is no doubt that this is a real-world problem and there are plenty more in sport, including methods of ranking individuals and teams. It would be important though to use contexts that are relevant to the learners; for example, in Jamaica, clearly the use of mathematics and statistics in the area of athletics would be highly relevant.

Number of paths

Start from the top left corner of the field represented in Figure 12.3 and walk to the bottom right corner of the field, using the grid lines (paths between nodes) and moving only down or to the right.

> **Problem:** *How many different routes can you find?*

This appears to be an easy problem but learners, whatever their age, soon realise that it is not straightforward. Younger learners will want to draw out all the possibilities, but this is not an easy task. Even the more experienced

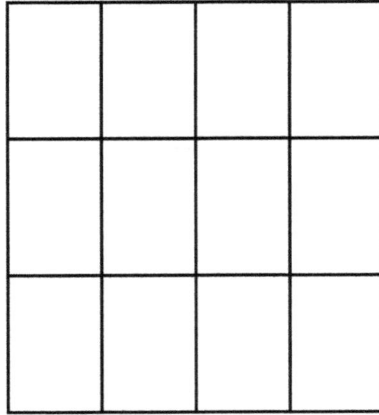

Figure 12.3 Grid lines representing paths and nodes.

learners often have difficulty in making progress; one plan is to look at more simple networks first, for example, a 2 × 2 network. As a general strategy, looking at a simpler but similar problem can sometimes be a starting point for dealing with a more complex example.

In Figure 12.4, we can see one way of moving to a solution: look at the first node that you can reach and note the number of routes to reach it. You can continue in this way, for each node noting the number of ways to get to it. In this way, you will see that the answer is 35 and, at a higher level, you can also spot that the Binomial coefficients are appearing on the diagonals.

Open problem solving

In *The Teaching Gap*, Stigler and Hiebert (1999) report that in all the countries studied, problem solving is seen as an essential part of the mathematics curriculum. Their findings also indicate that in all high achieving countries, problem solving was used to actually *drive* learning not merely to test learning.

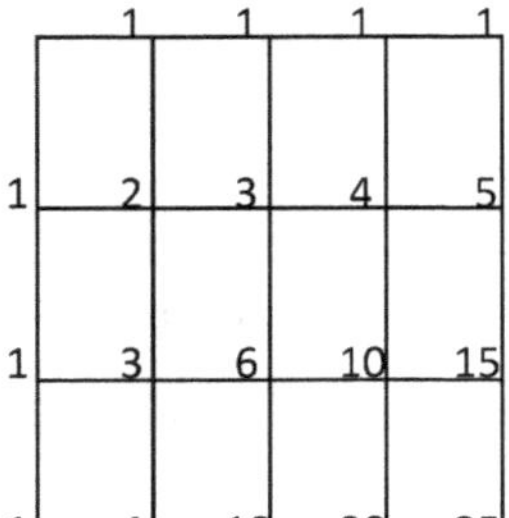

Figure 12.4 Grid showing number of moves to reach each node.

They suggest that it is in the choice of problem and the thinking behind this choice that cultural differences begin to show. Problems are carefully chosen not only to help learners develop their understanding of a particular concept but also to capture their interest by making the tasks thought provoking.

Stigler and Hiebert emphasise that in Japan, the "teacher presents a problem to the students without first demonstrating how to solve the problem". In this way, learning begins with a problem to be solved, and the problem is posed in such a way that learners need to gain new knowledge before they can solve the problem.

Much of the current Japanese approach to problem solving can be traced back to the research carried out between 1971 and 1976 by Japanese researchers in a series of projects on methods of evaluating higher-order thinking in mathematics education which became known as the "Open Approach" (Shimada, 1997).

The aim of open approach teaching is to develop both the creative activities of the learners and their mathematical thinking simultaneously through problem solving. It is based on the belief that learners' perceptions of mathematics are formed by the work they are asked to do. If they are mainly asked to carry out pre-taught procedures in a set of exercises, they will think mathematics is about following a set of rules. If we want them to think that mathematics is about solving problems, then they need to spend most of their time solving problems (Hiebert et al., 1997).

When open problems are used in mathematics teaching, learners have the opportunity to behave like creative mathematicians. In reality, to understand "openness" in Japanese problem solving means rethinking the way we teach mathematics. This is a crucial point as most teachers of mathematics rely on the "transmission" methodology in which teachers transmit the knowledge through instruction and then it is reinforced through practice. The strategy for "open problem solving" requires an almost directly opposite approach but the potential gains in learning are immensely valuable in producing young learners who are mathematical thinkers.

The benefits for the learners are clear but will only be effective with teachers who understand the model, are themselves capable and confident mathematicians and are willing to take risks. It is also essential to have problems that motivate both the teacher and their learners, that start where the learners are and have the potential to move their mathematical knowledge onward.

Mathematics in the real world

A common reaction from so many learners is "What is the use of this?" when faced with topics in mathematics – and for teachers it is not easy to justify. Although mathematics and in particular, mathematical thinking, is a key attribute for many young employees, to bring concepts to life with real world examples is not easy but it is important that problems are seen to be relevant.

Mathematics is so often seen by learners as a topic they are forced to take throughout their school lives and, whilst charismatic teachers can make it lively, for example, by using open problem solving, learners in secondary education need more justification and an appreciation that the mathematical topics and understanding really are important for later life and employment.

In post-16 education, there are now attempts to bridge this gap between maths in the real world and maths lessons in school and colleges. In part, this hinges on an appreciation of the idea of mathematical modelling as shown in Figure 12.5.

This illustrates the idea that the real-world problem is turned into a mathematical problem by making assumptions that simulate a real problem.

This is solved and the outcomes used to either explain or predict or to look again at the assumptions made and go round the cycle again.

Here are a two problem-solving activities based on mathematical models.

Heptathlon points formula

In the Woman's Heptathlon in athletics, there are seven activities, the last one being the 800 m running race.

Each event is scored separately: world class athletes are expected to score around 1000 points in each event. Each event has a formula similar to the one given here for the 800 m:

$$P = 0.11193(254.00 - T)^{1.88}$$

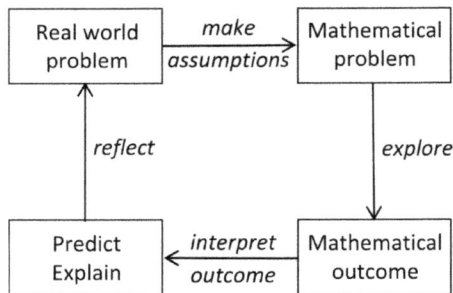

Figure 12.5 Mathematical modelling cycle.

Investigate these problems:

- *What times scores zero points?*
- *What time scores 1000 points?*
- *Is this scoring system fit for purpose?*

This is a precise mathematical model designed to provide points for this event; note that the other six events all have similar models. These models were designed a number of decades ago being based on previous data from expert athletes. As they are used for all international events, including the Olympic Games, it could be assumed that they are 'fit for purpose'.

More recent data seem to indicate that for some of the events, such as the shot put and javelin throw, it appears to be more difficult to reach the expected 1000 points compared with the running and jumping events. It has been suggested that this might be due to substances taken by athletes at the time when these models were designed.

Another downside is that most people will not understand the model, let alone be able to calculate the points scored! So a simpler model could, for example, be of the form:

$$P = \frac{(240 - T)^2}{9}$$

The use of mathematical models in sport is an important area of application and whilst this might only appeal to some learners, it is easy to understand and appreciate. Examples include:

- World rankings for individuals in, for example, tennis and golf;
- World rankings for countries in, for example, cricket and football.

Sport is certainly an area where mathematics can be seen to be used in order to solve a specified problem and for many learners, it is a motivating and illuminating topic; they can appreciate the problems and are keen to discuss the issues.

Optimum ferry loadings

In the Outer Hebrides in Scotland a ferry carries vehicles between the Isle of Harris and North Uist. The car deck of the ferry has three lanes, each of 20 m length. The vehicles awaiting loading are listed in Table 12.2.

Problem: *Can all the vehicles be carried on this ferry?*
Extension: *Can you design a method of approach for solving problems of this type, given that there are n lanes each of length l, and a variety of vehicles of known length to be carried?*

Table 12.2 Length of vehicles to be loaded

Vehicle	Length (m)
Car	3.5
Car	3.0
Van	4.0
Estate car	4.5
Furniture van	10.5
Milk tanker	14.0
Car and boat trailer	8.5
Breakdown van and car	10.5

This is a practical problem that is pertinent to so many industries and, whilst this is the one-dimensional version, there many similar problems in two or three dimensions, for example:

- optimum loading of shipping containers or cargo aeroplanes;
- cutting out nets for shapes.

This comes under the general theme of "bin-filling" and perhaps surprisingly there is no algorithm to find the optimum solution to this problem but many methods get close to the optimum solution.

For learners though, it is an easy problem to understand and to think of procedures to work towards an optimum solution.

Key priorities in mathematics education for developing countries

For many teachers using the open approach (including examples of mathematical modelling) takes them "out of their comfort zone". However, it is important, indeed vital, that learners are provided with an enhanced mathematical curriculum that aims not just to teach topics with practice and applications but to help them become mathematical thinkers.

Tomorrow's world is likely to require mathematical thinkers who have the ability and confidence to use and apply their mathematical skills and knowledge when faced with unseen problems; that is, problems that will need transferable skills. Teaching using the open approach has to be a key component of continuing professional development of teachers, both primary and secondary. This is our first recommendation:

Continual professional development for all teachers of mathematics should focus on teaching and learning through problem solving using both external and local school-based experts, using Japanese-style lesson study to help and support teachers to enhance teaching and learning in their classroom.

We highlight here two more priority areas for enhancing mathematical education in developing countries, recognising that change in teaching strategies takes time (and support) and that countries that have good performances, such as Finland and New Zealand, in the secondary sector have achieved this in part by investing in their teachers in the primary sector, ensuring that their learners gain aspects of mathematical thinking at an early age.

So the **next priority area** is the **primary stage of education**. It is vital that young learners become confident in the use and application of number; that is they gain a sense of number so that they feel confident and capable when using numbers as well as beginning to understand early algebraic thinking.

So we recommend that:

Teachers in the early years of education should be encouraged to

- *Help learners to develop a deeper understanding of the concepts that underpin mathematics;*
- *Provide extra practice on key topics, particularly in early number work;*
- *Encourage problem-solving activities and extension questions.*

To achieve these aims, we suggest the adoption of recommendations listed as follows which have been adapted from Enhancing Primary Mathematics Teaching and Learning (Educational Development Trust, 2012).

1 **Lessons** are well prepared, board prepared in advance, learners with resources needed.
2 **Seating:** learners have eye contact with teacher and can demonstrate answers at the board.
3 **Whole–class interactive teaching** predominates, with planned intervals of individual and paired work. All learners on task and demonstrate, answer and explain to the class.
4 **Friendly, non-confrontational atmosphere** where learners learn from and support others and have fun! Mistakes used as teaching points. Encouragement given to those who have difficulty and praise given when deserved.
5 **Exercises** reviewed interactively with the whole class at the time. Learners give the solutions, not the teacher, and the rest of the class agrees/disagrees or suggests alternative solutions.
6 **Problems, challenges** or **extension** work set for able learners, or they help less able neighbours; no one is inactive and extra practice for learners who require extra help.
7 **Introductory** and **reinforcement** tasks to ensure no learner is left behind.

8 **Correct notation, layout** and **language** used at all times. Teacher acts as a model for learners to follow (on board and orally).
9 **Good pace** and **varied activities** related to the concept being taught. Time limits set for individual/paired work. Time allowed for learners to explain and for whole class discussion.

We also recommend that teachers at all stages including teachers in early years education should and need to have a sound **mathematics subject knowledge** and this is another aspect that requires investment both in time (continuing professional development) and money (experts to help).

Our **third priority area** is in the upper **Secondary stage of education**. Here it is so important to keep the interest of learners and promote the skills needed for future employment. For highly academic learners, this is probably not a problem but for others it is crucial to keep their enthusiasm and motivation for mathematics. One way to achieve this is to adopt a problem-solving strategy, with an emphasis on problems in context.

In Japan, Sawada (1997) emphasises that a suitable problem for learning mathematics should have the following features.

The problem must begin where the learners are

The design or selection of the task should take into consideration the current understanding of the learners. They should have appropriate ideas to engage and solve the problem and yet still find it challenging and interesting. In other words, it should be within their zone of proximal development.

The problematic or engaging aspect of the problem must be due to the mathematics that the learners are to learn

In solving the problem or doing the activity, learners should be concerned with primarily making sense of the mathematics involved and therefore developing their understanding of those ideas. Although it is acceptable and even desirable to have contexts or external conditions for problems that make them interesting, these aspects should not overshadow the mathematics to be learned.

The problem must require justifications and explanations for answers and methods

Learners should understand that the responsibility for determining if answers are correct, and why, rests with them. Learners should also expect to explain their solution methods as a natural part of solving problems.

In this way, teaching with problem-based tasks is learner centred rather than teacher centred. It begins with and builds on the ideas our learners have available – their workings and understanding. It is a process that requires faith in learners and a belief that all learners can create meaningful ideas about mathematics.

This methodology underpins the development of Core Maths courses (Department for Education, 2015) for post-16 education in England, where courses are designed with the following objectives:

Objective 1: Deepen competence in the selection and use of mathematical methods and techniques;

Objective 2: Develop confidence in representing and analysing authentic situations mathematically and in applying mathematics to address related questions and issues;

Objective 3: Build skills in mathematical thinking, reasoning and communication.

There is reduced emphasis on higher order mathematical topics but more on how learners select and apply their mathematical knowledge to problems in meaningful contexts. This is a relatively new development for England but we think significant in terms of providing learners with the necessary **transferable skills** where the problems and issues of tomorrow will not be the same as those of today (see NCFE, 2018). Courses similar to this, with the use of appropriate contexts, could be highly applicable and motivating for learners in developing countries.

This leads to our recommendation:

At post-16 level, provide mathematics and statistics courses and examinations that show the relevance of mathematics through problems in contexts relevant and meaningful to the learners and teachers.

The digital world will continue to develop at pace and other international issues will arise. There will be new problems (and not just pandemic ones) that will need strategic solutions for developing countries. Developing countries will undoubtably need young employees with the ability to transfer their mathematical problem-solving skills to these new issues, whatever they are!

References

Department for Education. (2015). *Core maths qualifications: Technical guidance*. London, DfE. Available online at: https://www.gov.uk/government/publications/core-maths-qualifications-technical-guidance

Hiebert, J., Carpenter, T. P., Fennema, E., Fuson, K. C., Wearne, D., Murray, H., Olivier, A., & Human, P. (1997). *Making sense: Teaching and learning with understanding*. Portsmouth, NH, Heinemann.

Isoda, M., & Katagiri, S. (2012). *Mathematical thinking: How to develop it in the classroom*. Singapore, World Scientific Publishing.

Japanese Ministry of Education. (2008). Available online at http://www.globaledresources. com/products/books/guide_arithmetic_7-9.html

National Council of Teachers of Mathematics. (1980). *An agenda for action: Recommendations for school mathematics of the 1980s*. Reston, VA, Author.

National Council of Teachers of Mathematics. (1989). *Curriculum and evaluation standards for school mathematics*. Reston, VA, Author.

NCFE. (2018). New core maths qualifications. Available online at: https://www.ncfe.org. uk/blog/new-core-maths-qualifications

Educational Development Trust. (2012). Research report: Enhancing primary mathematics teaching and learning. Available online at: https://www.educationdevelopmenttrust. com/our-research-and-insights/research/enhancing-primary-mathematics-teaching-and-learning

National Research Council. 1989. *Everybody Counts: A Report to the Nation on the Future of Mathematics Education*. Washington, DC: The National Academies Press. https://doi. org/10.17226/1199

Nohda, N. (2000). Teaching by Open-approach method in Japanese mathematics classroom. In T. Nakahara & M. Koyama (Eds.), *Proceedings of the PME-24 conference* (Vol. 1, pp. 39–53). Japan, Hiroshima University.

Ofsted. (2011). Supplementary subject-specific guidance for mathematics. [Online]. Available at: http://www.ofsted.gov.uk/resources/generic-grade-descriptors-and-supplementary-subject-specific-guidance- for-inspectors-making-judgement.

Sawada, T. (1997). Developing lesson plans; in Becker, J. P., & Shimada, S. (Eds.). (1997). *The open-ended approach: a new proposal for teaching mathematics*. Reston, Virginia: National Council of Teachers of Mathematics.

Shimada, S. (1997). *The significance of an Open-ended approach*. In J. P. Becker & S. Shimada. *The Open-ended approach: A New proposal for teaching mathematics*. NCTM.

Singapore Ministry of Education. (2007). Available online at http://www.moe.gov.sg/education/syllabuses/sciences/files/maths-primary-2007.pdf

Stigler, J. W., & Hiebert, J. (1999). *The teaching gap: Best ideas from the world's teachers for improving education in the classroom*. New York, Free Press.

Chapter 13

The importance of data science for developing countries

Neville Davies

Introduction

We consider the new curriculum in data science (DS) created by the International Data Science in Schools Project (IDSSP) and recommend that policy makers and teachers of mathematics, computing and science in developing countries embrace this important new approach to learning from data. We will consider what it means to be statistically literate and suggest that developing countries embrace and promote data education in schools, colleges and universities at all levels.

Many definitions of *mathematical* literacy have been published over the years. For mathematics, the Organisation for Economic Co-operation and Development (OECD) defines mathematics literacy as follows:

> An individuals' capacity to formulate, employ and interpret mathematics in a variety of contexts. It includes reasoning mathematically and using mathematical concepts, procedures, facts and tools to describe, explain and predict phenomena. It assists individuals in recognising the role that mathematics plays in the world and to make the well-founded judgements and decisions needed by constructive, engaged and reflective citizens.

This statement forms the background to the OECD Programme for International Student Assessment (PISA) and, in particular, the test in mathematics (see OECD, 2016a, 2016b). This view of mathematical literacy refers to the role of the subject in the real world and that it can help in decision-making.

Unfortunately, it does not recognise the role mathematics plays in statistics and DS, the science of learning from data. DS, which many see as a development of the subject *statistics*, plays a key role in understanding and managing the world around us. We elaborate on this later in this chapter, where we highlight what it means to be *statistically* literate.

Historically, many schoolteachers in the United Kingdom at least, regard statistics as a branch of mathematics. In any case, many mathematics curricula contain aspects of statistics and its applications. The more progressive ones teach *statistical problem solving* as a way for school students to experience uncertainty and decision-making in the real world. In the next section, we consider statistical literacy, statistical problem solving and learning from data, and argue that teachers of mathematics in developing countries should give their students experience of solving real-world problems through embedding these topics in their own curricula.

Developing countries

The definition of a 'developing' country is not universally agreed. However, these countries are usually characterised by low achievement of industrialisation relative to their population size, and their standard of living, including income and education, is generally low. Indeed, an association exists between low income and high population growth. An *objective* definition usually involves using a country's gross domestic product (GDP) per head of population. However, the World Bank, through a series of blogs, see https://blogs.worldbank.org/opendata/should-we-continue-use-term-developing-world has downplayed the use of the word 'developing' claiming it has become less relevant and that it will phase out its use.

We note that, from the United Nations (UN) M49 standards there is a key reference to statistics in its comments about country–development descriptors:
The designations 'developed' and 'developing' are intended for statistical conven-ience and do not necessarily express a judgement about the stage reached by a par-ticular country or area in the development process (https://unstats.un.org/unsd/methodology/m49/).

Implicit in this comment are some very important ramifications. The improvement of poverty, education, living and other standards in all coun-tries depends on being able to *measure* a country's past and current status in these areas. Helping to improve all areas should be a country's aim, with specific goals in each.

Because of the variation and uncertainty associated with all measurements of data, it is necessary for each country to make use of *statistics* and *DS*. Usually statisticians or data scientists have the best skills and expertise to solve problems that involve making sense of data.

Irrespective of differences between definitions of 'developing countries', we contend that, owing to the rapid development of statistical methods for processing and analysing small to very large amounts of data, it is imperative that every country's national statistical system (NSS) has people who can contribute skills and knowledge to:

- formulate appropriate problems to be solved;
- measure and collect/produce data in a trustworthy way;
- analyse the data;
- communicate the results to policy makers, business, private individuals and the country's population.

Rather than import expertise from other countries, it would be much better for every developing country to educate and train its own people to gain skills and knowledge in the areas listed earlier – 'bread and butter' topics within statistics and DS. *Education* in statistics and DS should start in schools and run right through university and beyond. This will enable more and more people within countries to receive data education and become statistically literate, which, in turn, should benefit those countries. In addition, it will provide employment in an increasingly important area that is relevant in *all* countries, either developing or developed.

Statistical literacy and data science in Jamaica and developing countries

Statistical literacy is the ability to appreciate the art, science and practice within subjects where data (numbers in context, images or text) are used, needed or produced. Becoming statistically literate does not happen overnight; rather, it takes time for the ideas and concepts to mature, experience with applications to develop and will depend upon the individual's aptitude to appreciate that variability is all around us and impacts on everyone.

Statistical literacy uses a range of thinking and practical skills that include: knowledge; comprehension; application; analysis; synthesis; and evaluation. It enables a feel for data, including being able to support an argument with evidence, but also being aware of the variety of interpretations that are possible from those data. A statistically literate person will apply common sense to problems and appreciate that information that is gleaned from data will have uncertainty attached to it.

The statistical problem-solving cycle is a good example of statistical thinking in action and we suggest that getting practice in carrying out the cycle will help students become more statistically literate as it exercises many cognitive skills. See, for example, Marriott, Davies, and Gibson (2009).

Examples of problems that need statistical thinking to solve them include:

- deciding which mobile phone to buy;
- designing new drug trials;
- choosing between facial creams from different manufacturers;
- examining DNA evidence;
- analysing surveys to determine the most popular laptop;
- optimising the mass production of burgers;
- facial recognition in digital cameras;
- modelling the performance of fertilizer on crops;
- understanding risk and uncertainty.

Since problem solving in mathematics has become a key feature of the PISA tests, it may be a useful research project to identify those cognitive domains that are needed in various stages of solving *mathematical* problems. Posing a range of questions that focus on different cognitive domains might improve cognitive skills that, in turn, provide answers to help solve problems. In statistics, Marriott et al. (2009) showed that a surprisingly wide range of cognitive skills, as defined by the modification of Bloom's taxonomy of learning (Bloom et al (1956), by Anderson and Krathwohl (2001), are needed to successfully complete the statistical problem-solving cycle.

A characteristic that is common to all countries, whether fully developed, well developed or developing, is data. Data are the lifeblood of decision making at many levels, for governments and individuals. It is incumbent on educationalists and policy makers in all countries to help everyone to be able to understand what data can and cannot do in decision-making. Becoming data literate is a drip-feed process – many countries start the study of data, whether in mathematics curricular or in other subjects that use or produce data, from an early age. One way to study data is, of course, through the subject *statistics*, although there are other names used to describe the discipline, such as quantitative methods, analytics and more recently, DS. For a view on the relationship between statistics and DS, see Donoho (2017).

The IDSSP published a report in late 2019 that outlined an innovative approach to teaching and learning from data. An international team of educators from Mathematics, Statistics and Computer Science ran the project. The team was responding to a need for more and more people who can practice DS – demand is outstripping supply, worldwide. The IDSSP report is an initiative that makes innovative proposals for schools in *all* countries to enrich mathematics and other curricula with dynamic ways to learn from data, using recent advances in technology for data visualisation and analysis.

The potential for job creation in this area in developing countries appears to be very good. Furthermore, there is also a pressing need for all people to be able to go beyond the knowledge provided by mathematics and be more capable of 'understanding, interpreting, critiquing and making decisions based on data as they cope with the vagaries of life'. The initial, and primary, objective is that the IDSSP curriculum frameworks will provide the basis for development of courses in introductory DS for students in their final two years of secondary school, with corresponding courses to teach teachers how to teach introductory DS.

DS draws on several disciplines, including aspects of Computer Science, Mathematics and Statistics, together with areas such as problem elicitation and formulation, collaboration and communication skills. At the heart of DS is the *cycle* of learning from data. In fact, the activities for *doing* DS coincide with several of the recommendations made by the OECD in answering the ten questions in OECD (2016b). See the OECD questions 2, 8 and 9 and related comments.

The paradigm for carrying out DS is similar to the statistical problem-solving cycle that has been used in the United Kingdom for many years. Also known as the Problem, Plan, Data, Analyse, Conclude (PPDAC) cycle (Wild & Pfannkuch, 1999), Marriot et al. (2009) showed how the statistical problem-solving cycle could be taught, learned and assessed and related it to cognitive skills. The approach of teaching statistics through problem solving has been used in New Zealand, the United Kingdom, and the United States. In fact, the American Statistical Association (ASA) report *Pre-K-12 Guidelines for Assessment and Instruction in Statistics Education (GAISE) Report II* (Bargagliotti et al., 2020) recommends the use of the cycle in teaching statistics at all school levels. The following diagram, taken from IDSSP (2019), shows the cyclical nature of the DS paradigm (Figure 13.1).

Notice how each stage has a number of sub-stages that may need to be implemented before proceeding to the next stage in the cycle. It can be a continuous process, whereby an original problem may not be fully solved until some or all of the cycle is returned to after suitable discussion and communication. For example, we may need to collect more data, particularly if flaws in the original data-collection activity emerge.

We present a brief summary of the proposed curriculum framework in **Appendix 1**. More details are in IDSSP (2019) where it provides a curriculum

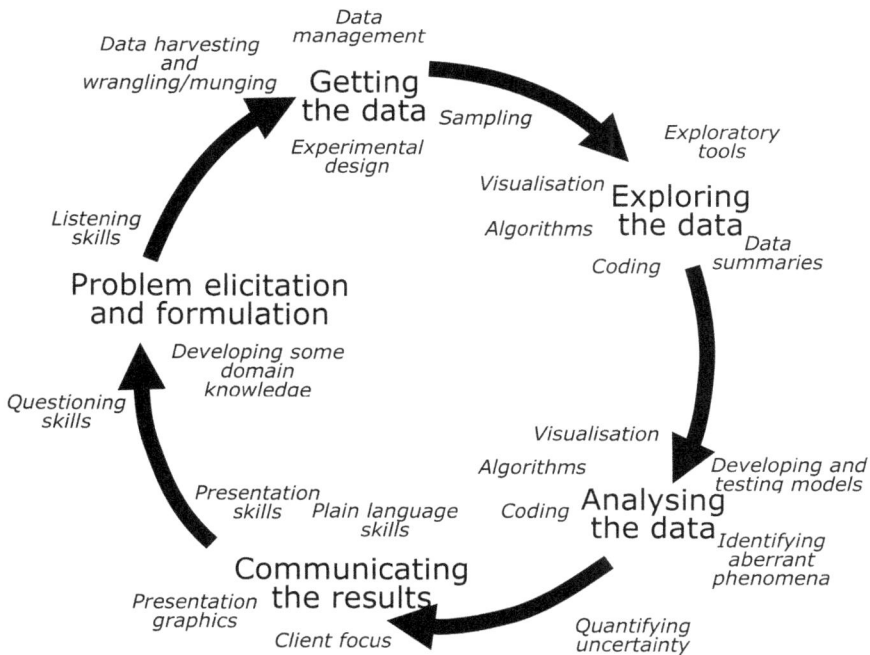

Figure 13.1 Data science cycle.

for teaching students and pedagogic material for teachers themselves to become knowledgeable in, and adept at, teaching DS.

The journal *Teaching Statistics (TS)*, established in 1979, is the only journal designed for statistics school teachers of students up to about age 19. Published by Wiley, on behalf of the TS Trust, it appears three times a year. In 2021, Wiley will publish a free-to-subscribers of *TS* an online e-book, entitled *Teaching Data Science and Statistics: Senior School or Introductory Tertiary*. In **Appendix 2**, we present brief details of intended content of the e-book. It will be available to buy for non-subscribers to the journal TS.

An interesting podcast in 2018 by Ofir Reich, a data scientist working at the Centre for Effective Global Action at UC Berkeley proposes that DS can help to end poverty in developing countries. (https://www.youtube.com/watch?v=ITeAvleCd2s&feature=youtu.be).

Examples of projects where Reich has worked for and on behalf of developing countries: identifying fake companies in India; enabling Afghanistan to pay its teachers electronically; and raising yields for Ethiopian farmers by messaging them when local conditions make it ideal to apply fertiliser. The podcast is very positive and provides a persuasive view of some roles for DS: we feel that educationalists and policy makers in so-called developing countries should take note and make great efforts to make the teaching and learning of DS in schools and beyond.

We recognise there may be one or two caveats in what we propose: there may be issues with the development and use of technology in developing countries. For example, modern methods of data collection use technology to collect small and large data sets. If a developing country has problems in this field, collecting good and trustworthy data may be difficult. Naturally, addressing these issues would be necessary in time for data scientists to be able to do their jobs efficiently.

Conclusions

The key conclusions are:

- In order to enhance the development of DS (including statistical literacy) in schools in developing countries, policy makers should embrace the recommendations of IDSSP (2019) and develop teaching and learning resources jointly with mathematics and computer science teachers.
- We recommend that teachers of mathematics, statistics and computer science in Jamaica and developing countries get together to discuss how DS in schools could benefit current and future generations of students to learn from data in a trustworthy and professional way.
- There may be some very useful and exciting employment opportunities for young people after studying DS. Students in developing countries

could join the ever-increasing teams of experts who are practising DS across the world.

- Developing countries may need to address other issues first, such as availability of technology for teaching and learning statistics and DS, as well as technology for carrying out the practical applications of the subjects as part of the job of their data scientists.

References

Anderson, L. W., & Krathwohl, D. W. (2001). *A taxonomy for learning, teaching, and assessing: A revision of Bloom's educational objectives.* New York, Longman.

Bloom, B. S., Englehart, M., Furst, E., Hill, W., & Krathwohl, D. R. (1956). *Taxonomy of educational objectives: The classification of educational goals, by a committee of college and university examiners. Handbook I: Cognitive domain.* New York, Longmans, Green.

Donoho, D. (2017). 50 Years of data science. *Journal of Computational and Graphical Statistics, 26*(4), 745–766.

Bargagliotti, A., Franklin, C., Arnold, P., Gould, R., Johnson, S., Perez, L., & Spangler, D. (2020). *Pre-K-12 guidelines for assessment and instruction in statistics education (GAISE) report II.* Alexandria, VA, American Statistical Association and Reston, National Council of Teachers of Mathematics. https://www.amstat.org/asa/education/Guidelines-for-Assessment-and-Instruction-in-Statistics-Education-Reports.aspx

International Data Science in Schools Project. (2019). Curriculum frameworks for introductory data science. http://idssp.org/files/IDSSP_Frameworks_1.0.pdf

Marriott, J. M., Davies, N., & Gibson, L. (2009). Teaching, learning and assessing statistical problem solving. Available online at: https://ww2.amstat.org/publications/jse/v17n1/marriott.html

OECD. (2016a). *How teachers teach and students learn: Successful strategies for school.* Paris, PISA, OECD Publishing. https://www.oecd-ilibrary.org/education/how-teachers-teach-and-students-learn_5jm29kpt0xxx-en

OECD. (2016b). *Ten questions for mathematics teachers … and how PISA can help answer them.* Paris, PISA, OECD Publishing. https://dx.doi.or/10.1787/9789264265387-en

Wild, C. J., & Pfannkuch, M. (1999). Statistical thinking in empirical enquiry. *International Statistical Review, 67*(3), 223–248.

Appendix 1: Curriculum framework for data science

The documents provided by IDSSP (2019) provide the following motivation for development of the curriculum.

'The purpose of this international collaborative project is to transform the way teaching and learning about Data Science is carried out in the last two years of school, with two objectives:

1 To ensure that school students acquire a sufficient understanding and appreciation of how data can be acquired and used to make decisions so that they can make informed judgments in their daily lives, as students and then as adults. In particular, we envisage future generations of lawyers, journalists, historians and many others, leaving school with a basic understanding of how to work with data to make decisions in the presence of uncertainty, and how to interpret quantitative information presented to them in the course of their professional and personal activities.

2 To instil in more scientifically able school students sufficient interest and enthusiasm for Data Science that they will seek to pursue tertiary studies in Data Science with a view of making a career in the area'.

We reproduce the curriculum proposals and suggest the reader study these and scrutinise the document produced by IDSSP (2019).

Curriculum proposals

1 Introductory Data Science

1.1   Data Science and Me

Aims: To help students become aware of the importance of data in their lives, the exciting possibilities opened up by Data Science, some of the big ideas of Data Science, related social and ethical issues and to introduce students to the Data Science learning cycle.

1.2   Basic techniques for exploration and analysis (BTEA).
Part 1: Tools for a single feature/variable
Part 2: Pairs of features/variables
Part 3: Three or more features/variables

Aims: to *introduce* simple data-storage formats, graphical displays and numerical summaries that are useful in their own right; to provide a basis for building on in many subsequent units and to give students experience with using these tools to make discoveries in data.

1.3   Graphical displays and tables

Aims: to develop the students' understanding of appropriate choices and uses for graphical displays and tables in learning from data and when presenting the results of an analysis.

1.4   The data-handling pipeline

Aims: to introduce the tools used to deal with data (sometimes called data wrangling), and to develop an understanding of data management issues in the context of the Data Science learning cycle.

1.5   Avoiding being misled by data

Aims: to provide students with a deeper understanding of how to critique data and data-based claims, including an appreciation of the ideas of bias, confounding and random error;

- to introduce them to some good practices for obtaining reliable data (random sampling and randomized experiments);
- to motivate incorporation of uncertainties in estimation using margins of error or interval estimates;
- and to provide some introductory experience with the ideas of statistical testing in the context of an experiment for comparing two treatments.

2 Further Data Science

(Select from these stand-alone topics, but 2.4–2.5 and 2.8–2.9 are core to DS thinking)

2.1   Time series data

Aims: to develop basic understanding and skill in displaying, exploring, interpreting and presenting results for data that take the form of a time series.

2.2   Map data

Aims: plotting (positional or regional) geo-located data plotted on maps, use for exploratory analysis and understanding maps themselves as graphical representations.

2.3   Text data

Aims: to appreciate the many contexts in which text data can arise, to learn to explore such data and to extract and present potentially interesting characteristics in practical settings.

2.4   Machine learning: Supervised

Aims: to develop an understanding of some of the contexts in which classification and prediction problems can arise, and to learn how to apply some basic tools for classification and prediction to draw conclusions in practical settings.

2.5   Machine learning: Unsupervised

Aims: to develop an understanding of some of the contexts in it is of interest to find groups in data ('cluster analysis'), and to learn to apply some basic tools for this purpose and present the results in an informative fashion.

2.6 Recommender systems

Aims: to learn about some of situations in which Recommender systems are used, the sorts of data that are collected to develop these systems and some methods for building such systems.

2.7 Interactive visualisation

Aims: to learn how interactive visualisation can be used to enhance various steps in the Learning from Data cycle, particularly relating to exploring data and communicating results and to gain skills and experience in applying some of the basic tools.

2.8 Confidence intervals and the bootstrap

Aims: An introduction to important concepts of confidence intervals in a random sampling context implemented using simulation methods (bootstrap resampling).

2.9 Randomisation tests and significance testing

Aims: An introduction to important concepts of significance testing in the context of randomised experiments implemented using simulation methods (randomisation/permutation tests).

2.10 Image data

Aims: to appreciate the many contexts in which image data can arise, to learn to explore such data and to extract and present potentially interesting characteristics in practical settings.

Appendix 2: Aim and objectives of e-book

'Teaching Data Science and Statistics: Senior School or Introductory Tertiary'

The overall aim of the e-book is to be useful to all who are actively involved in data education at the senior school and/or introductory tertiary levels, whether in teaching, curricula and resource development, curricula implementation or policy development and implementation.

Because the range of influences of data science and statistics is so great, chapter objectives will include a clear specification of audience. Chapters will fall in one of the following categories.

• For teachers: bringing data science influences, approaches and skills into existing senior school curricula.
• For senior school curriculum developers and educational authorities: developing data science curricula and/or incorporating data science within statistics at senior school level.

- For tertiary teachers: incorporating data science approaches and skills within introductory statistics programs, including across disciplines.
- For tertiary curriculum writers (often those who will also teach the curricula) and program developers: data science curricula integrating statistics, and statistics integrating data science.
- For university authorities: specifying how data science should be taught in and across university programs.

Chapter 14

Summary of evidence and recommendations

David Burghes and Jodie Hunter

Introduction

It is evident from a review of the chapters in this monograph, that there are some key findings related to developing mathematics education for sustainable economic growth and job creation. Also evident is that there are no easy answers: however, it is apparent that, for the future economic well-being of developing countries, investment in education is paramount. In our current context, with much of the world hugely affected by the COVID-19 pandemic, economies worldwide being decimated and countries going into severe recession, it is more important than ever to invest in the education of young people. There will be real opportunities for developing countries to make significant progress as the established world order can potentially change significantly. Many countries with established flourishing economies have been hard-hit by the virus and will take decades to recover.

It is clear, though, that as the world recovers, we will need robust and equitable educational systems, with mathematics education being of crucial importance for the next generation of young people. Events such as the Maths Summit hosted by The Mico University College and initiated by Professor Neville Ying play an important role in facilitating the sharing of best practice from around the world in relation to mathematics education. As described in the introductory chapter, 'Setting the Scene' by Ying, there was no thought of international pandemics when planning the Summit, but his vision and priorities for job creation are even more relevant today. In part, this book contributes to a wake-up call for mathematics education as we move into the next industrial revolution.

Mathematical achievement and evidence from mathematically high performing countries

Programmes of international testing studies such as PISA (Programme for International Student Assessment) and TIMSS (Trends in International Mathematics and Science Study) offer those in mathematics education opportunities to examine different educational systems, pedagogy and practices. However, it is also important to consider both the relevance and suitability of developing countries participating in international programmes such as PISA. As Davies and Georgeson (Chapter 2) explain, there are a number of implications for developing countries taking part in PISA with some benefits for both teachers and policy makers. However, a key finding is that it is important for policy makers and educators not to focus unduly on the rankings as there is statistical uncertainty about the whole process of international comparisons. Rather than focusing on rankings PISA reports can be used beneficially to delve into the wealth of information that they provide related to the different types of pedagogy used in participating countries.

Drawing on the framing of PISA as an opportunity to investigate pedagogy brings us to looking in particular at practice in established countries that do well consistently in international rankings. It is important to note that these may include contrasting educational systems such as highlighted in this book with chapters focused on both Finland, from the region of Europe (Chapter 3) and Japan, in the region of Asia Pacific (Chapter 4). Both Finland and Japan were developing countries following the Second World War and both demonstrate examples of economic growth and job creation while providing high-quality mathematics teaching.

As highlighted by both Pahkin and Koyama, despite key differences in their educational systems, Finland and Japan strive for continual improvement regarding mathematics education, and reform their educational systems as they deem necessary. Pahkin (Chapter 3) provides a summary of key aspects of the Finnish education system which can be linked to best practice for developing countries. This includes moving to a more equitable educational system by replacing a two-tier system of grammar schools and civic schools with a comprehensive school system, which is free throughout to all, up to and including degree level. Interestingly, the Finnish education system avoids standardised testing while maintaining the basics of mathematics as a priority. Other elements include pedagogical practices focused on teaching students to use mathematical thinking, and a highly qualified teacher workforce.

In Chapter 4, Koyama focuses on mathematics education in Japan and provides a detailed account of how problem solving and the professional development process of lesson study has improved mathematical performance. A key message from Koyama based on his work in Japan, is that teaching mathematics through problem solving is a highly successful method of enabling students to think mathematically either individually or in groups. We see in his chapter how teachers can use *lesson study* to help them adopt a problem-solving approach to teaching mathematics. Importantly, the key components of lesson study to improve mathematics teaching and learning are the collaborative study of teaching materials, group development of a research lesson plan, lesson implementation and group observations and reflective analysis. The process of *lesson study* could potentially be adopted in developing countries, as this is a collaborative and continuous form of professional development for teachers.

Considering innovations in mathematics teaching and learning for developing countries

An important aspect of effective and equitable mathematical education systems is considering innovative practice which caters for *all* learners, including those from diverse backgrounds. This is of relevance to not only developing countries but also to all countries around the world. In this book, we have drawn on examples from the Southern Hemisphere which also undoubtedly have wider applications. Two of these chapters (Chapters 5 and 6) have a focus on Pāsifika people both in the Pacific and within New Zealand. A key argument made in both chapters is the need to consider how we can change mathematics pedagogy and practice to align with diverse learners' cultural experiences and values. Hunter and Restani (Chapter 5) provide evidence that, when teaching children from Niue, a small self-governing island nation in the Pacific, mathematics contexts, examples and tasks should align with the home and family experiences of those children. Using photography and photo-elicitation interviews, the authors highlight the rich range of contexts in the children's lives that can be authentically linked to mathematics. Consequently, they argue that in developing a curriculum, resources and mathematical tasks, developers should consider the existing *funds of knowledge* of learners.

This is particularly pertinent in developing countries where groups of learners have unique experiences and funds of knowledge. A key argument from Hunter and Restani is that such an approach can support achievement that is more equitable because learners can access challenging, problematic tasks without struggling to make sense of the context. Secondly, they make the point that links between home and school contexts are imperative if we want learners with alternative funds of knowledge to see the relevance and importance of mathematics, and to continue studying the subject throughout their school lives.

In Chapter 6, Hunter critically analyses how pedagogical practices in mathematics classrooms are required to shift in order to encompass and build on the cultural beliefs of diverse learners. She uses the example of Pāsifika people to show how their cultural values are constructed within, and shaped by, their collectivist and communal ways of life. Within the New Zealand context, this contrasts with a focus on competition and individualism in schooling and as a result, many become underachievers across many subjects and specifically within mathematics. For example, the practice of placing Pāsifika students in ability groups conflicts with the beliefs and values of many Pacific Nations people because these grouping structures encourage competitiveness and place importance on the individual rather than *group* success. Enhanced engagement for Pāsifika students entails the use of more open and flexible pedagogy built around teachers noticing and responding to student reasoning and participation. The teachers needed to develop a safe and supportive learning environment, making direct links to their students' family contexts and emphasising that family members take different roles, all of which collectively make different cultural activities successful. From this chapter, we can see the need to consider wider aspects of mathematics education including culture and identity when reflecting on how to develop mathematics education that results in sustainable economic growth and job creation.

Another important component of equitable mathematics classrooms for diverse students is highlighted in Chapter 7 by Hill, Kern, van Driel and Seah. They contend that within mathematics education, there is often a pre-occupation with the development of cognitive skills and raising academic achievement, whilst discounting the importance of student well-being. Instead, these researchers argue, we need to consider well-being dimensions because when these are applied to mathematic education there are links to positive mathematics learning outcomes. The data in their chapter draws on responses from a large number of students from different ethnic backgrounds living in Australia. Importantly, for all students, positive classroom relationships was by far the most common dimension, followed by a sense of engagement, mathematical understandings and cognitive accomplishments, positive emotions, perseverance, music and meaningful mathematics learning.

There were more consistencies than differences across ethnicities, suggesting that the general well-being dimensions identified in developed countries may translate well to developing countries. Students referenced a 'good' teacher as one who is engaging or fun, who makes mathematical explanations with clarity, a teacher who understands the learning needs of his/her students and provides individualised assistance when required. The key findings from this study highlight that there is more to mathematics education than merely developing cognitive skills and achieving academic benchmarks. In all countries, we need to consider that student mathematical well-being also matters and reflect on how mathematics teaching can address this.

Developing mathematics education systems that address sustainable economic growth and job creation also requires a highly skilled mathematics education workforce. Within this book, we highlight different frameworks that can be used to upskill the mathematics education practitioners. These include consideration of how to develop pedagogical practices such as those illustrated by Seshaiyer (Chapter 8) and how to address mathematical content knowledge with teachers in Chapter 9 by Binns-Thompson. In his chapter (Chapter 8), Seshaiyer describes some novel educational frameworks that provide the opportunity for mathematics educators to not only engage students through mathematical tools to represent, understand and solve real-world problems, but also to engage them in using tools when making a decision, prediction or solution involving a real-world problem.

These frameworks help to prepare students to become life-long learners who can then go on to pursue state-of-the-art jobs. He makes the point that to create this next generation of students, however, we must upskill our current educators with pedagogical practices that go beyond just delivering content. Specifically, mathematics education must include a variety of learning approaches including experiential learning, inquiry-based learning, challenge-based learning and interdisciplinary problem-based learning. At the centre of innovative pedagogical practices to advance mathematics teaching and learning in the 21st century is authentic mathematical tasks. Importantly, this links to the solving of real-world mathematics problems, an important recurring theme in this book.

Implicit across many chapters within this book is the importance of a highly skilled mathematics education workforce. As Binns-Thompson (Chapter 9) highlights, adequate subject knowledge of teachers is required to provide interesting and relevant lessons that help to enhance the mathematical progress of their learners. She argues that although enhanced mathematical knowledge is not a guarantee of effective teaching of mathematics, it remains a necessary condition for effective teaching. Specifically, mathematics content knowledge enables a teacher to understand misconceptions of learners, provide alternative strategies to support understanding, facilitate learners to generalise through extending topics, motivate and progress learning, confidently and capably pursue different approaches and solutions by learners and help learners become mathematical thinkers. Chapter 9 draws on data from a developing country, Jamaica, and reports on the impact of providing a subject knowledge enhancement platform (SKE) for Jamaican teachers of mathematics.

Overall, participants in the study enjoyed the challenge and were successful in enhancing their mathematical knowledge. However, it is also important to consider barriers in developing countries that may make participation in such innovations difficult. In the case highlighted in Chapter 9, this included teachers' fear of mathematics, limited access outside school to the required technology, and time constraints for busy working teachers. Even though enhancing mathematical knowledge is a long-term strategy for success, it is a vital one for enhanced mathematical progress in developing countries. Providing an effective mathematical foundation in learners' early years gives a basis that can be built on later in schooling. Effective and strong schooling systems then lead to strong economic growth through the development of an educated and innovative workforce.

A key aspect of mathematics education required to support modern 21st-century economies is the development of science, technology, engineering and mathematics (STEM). Benjamin and Baker-Gibson (Chapter 10) consider the implications of the STEM agenda for developing countries and

Jamaica in particular. They illustrate that higher education offerings within the Caribbean nation states have not kept pace with the required output of STEM professionals. Importantly, they note the following difficulties: firstly, many of the Caribbean workforce will lack essential mathematical skills and competencies and secondly, there is an inadequate number of trained mathematics and science teachers and in particular, inadequate access to quality science education. In contrast, the authors look to countries such as South Korea, China and Singapore, where technological innovation has demonstrated significant promise for national growth and development in economies. These countries rank consistently well in the PISA Mathematics and Science assessment.

It is important to note that Jamaica and other countries in the Caribbean do not lack mature educations systems, but there is an absence of country level or regional coordination designed to optimise outcomes in the STEM subject areas. A key recommendation for Jamaica and countries in the region, along with developing nations and regions around the world, is the need for consideration of a national STEM ecosystem as a basis for increasing the number of STEM professionals in support of overall innovation for economic development.

Priorities for future actions and final thoughts

When considering mathematics education for sustainable economic growth and job creation, it is important to consider the priorities for future action. Throughout this book, we have provided examples of best practice and key findings related to effective mathematics teaching. All of these should be considered as future priorities along with the two important approaches highlighted by Robinson (Chapter 11), Burghes and Sani (Chapter 12) and Davies (Chapter 13).

Firstly, in Chapter 11, Robinson describes how Lesson Study (introduced in Chapter 4) is a powerful CPD initiative to develop and enhance teaching and learning in countries beyond Japan. He clarifies the key procedures to be followed to ensure its effectiveness and illustrates this in the context of teaching mathematics through a focus on problem solving. He describes in detail how lesson study can be an effective method of generating, accumulating and sharing professional knowledge. This should a key initiative for developing countries to enhance and sustain great teaching and learning.

A key aim in mathematics education should be to develop lessons that motivate learners and enhance their learning. Burghes and Sani, in Chapter 12, advocate problem solving to enhance lessons. They contend that problem-solving activities can be used to motivate and encourage learners and support them to work collaboratively to find suitable solutions. Similar to what was advocated by Koyama (Chapter 4), they highlight how open problem solving, where only the problem is posed and not the strategies for solving, provides the learners with opportunities to explore and think 'mathematically'.

Using this approach, which can be seen to have advantages in learner engagement, requires teachers to challenge themselves and step away from normal teaching practices, paralleling the challenges Hunter describes in Chapter 6. Key to this is the use of real-world problems or at least contexts in the real world that are interesting and can be understood by the learners. Again, this links to the approach described by Hunter and Restani in Chapter 5.

Burghes and Sani (Chapter 12), describe the development of the new 'Core Maths' courses in England to motivate and encourage learners to use their mathematical and statistical skills in new areas of applications. They see that these 'applied' courses with the focus on problem solving rather than higher level mathematics are crucial for young people for future employment. Another key area for investment is that of primary mathematics and, in particular, early years. For the future prosperity of a country, the early years of education are the time when a mathematical (not just numerical) foundation can be put in place but this does need teachers who have a real understanding of mathematics and what lies ahead for their learners.

Similarly, of importance for developing countries is the development of Data Science. In Chapter 13, Davies makes the case for a new curriculum in Data Science created by the International Data Science in Schools Project and recommends that policy makers and teachers of mathematics, computing and science embrace this important new approach to learning from data. Elements of this include consideration of what it means to be statistically literate and the suggestion that developing countries embrace and promote Data Science education in schools at all levels.

This would benefit job creation and ensure that new positions can be filled by the local population rather than importing personnel from other countries. In order to enhance the development of Data Science and statistical literacy in schools in developing countries, policy makers should consider the recommendations of IDSSP (2019) and develop teaching and learning resources jointly with mathematics and computer science teachers, very much in line with the STEM agenda.

Throughout this book, we note many themes and findings that highlight key aspects of effective mathematics teaching which contribute to sustainable economic growth and job creation. Through both the Maths Summit in Kingston and the chapters in this book, we highlight many of the wide-ranging efforts of the mathematics education community to find and evaluate good practice relevant to countries with developing economies. We hope that you will gain an insight into the many suggestions and recommendations being put forward. These include, in no particular order:

- Cooperation not competition between schools
- No national testing and reporting

- Free education for all
- Problem solving
- Lesson study for CPD
- Culturally sustaining pedagogy including funds of knowledge
- Well-being of learners
- Novel frameworks for real-world problems
- SKE
- STEM ecosystem
- Focus on early years
- Problem solving in the real world
- Data Science cycle.

We are sure this list can be extended but I think our final words should be to thank our authors for their insights and to encourage developing countries to invest in education, in particular in mathematics and statistical education, for their future economic well-being.

Index

For Product Safety Concerns and Information please contact our EU
representative GPSR@taylorandfrancis.com
Taylor & Francis Verlag GmbH, Kaufingerstraße 24, 80331 München, Germany

* 9 7 8 0 3 6 7 5 0 0 3 0 6 *